AF900090

Performance Research

Performance Research is an independent, peer-reviewed journal published by Routledge Journals, Taylor & Francis Ltd for ARC, a division of the Centre for Performance Research Ltd, an educational charity limited by guarantee.

Performance Research was founded in 1995 by Ric Allsopp, Richard Gough and Claire MacDonald.

www.performance-research.net

CENTRE FOR PERFORMANCE RESEARCH

PRIFYSGOL ABERYSTWYTH UNIVERSITY

Forthcoming Issues

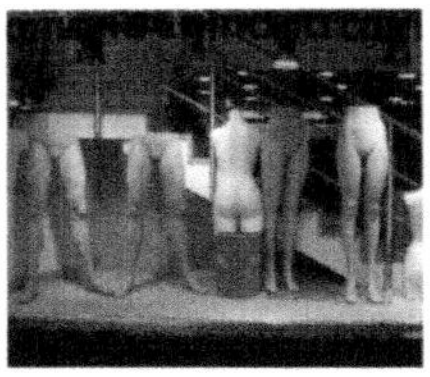

on Appearance
Vol.13 No.4

Beginning from the conviction that appearance matters – and matters as the very 'stuff' and substance of the kind of things we call performance – this issue examines the materiality of appearance as a key component of theatrical and social events. Exploring the role appearance plays in a range of cultural forms – from body art to live TV, shamanic invocation to video installation, magic show to 'non-professional' performance – *On Appearance* charts the construction, circulation and contestation of some of the imagined possibilities, lived realities, political identifications, and performative opportunities opened up by thinking through the logic of appearance. As well as examining the correlation between modes of appearance and practices of disappearance, and investigating their inscription in the recuperative dynamics of power, *On Appearance* explicates the ways in which appearance matters in affecting and positively producing the conditions, forms and relations structuring what Jacques Rancière calls 'the distribution of the sensible': the political organisation of sense-making activities within the intelligible framework of the visible.

Performing Literatures
Vol.14 No.1

The relationship between text and performance has often been seen as pivotal to the notional division between theatre and live art. If theatre is drama is literature, performance is performer is liveness. In the academy, with UK theatre departments emerging historically from literature departments, their propulsion away from over-reliance on text has led logically enough to a preoccupation with the *non-text-based*. But are any of these distinctions as neat as is often assumed? *Non-text-based* performance is often heavily reliant on text; while *literary* drama is frequently more concerned with the live and processual than many would allow. *Performing Literatures* takes a fresh look at the academic and professional landscape, attempting to reconsider the text/performance dichotomy from a variety of critical perspectives.

Submissions

Performance Research welcomes responses to the ideas and issues it raises. Submissions and proposals do not have to relate to issue themes. We actively seek submission in any area of performance research, practice and scholarship from artists, scholars, curators and critics. As well as substantial essays, interview, reviews and documentation we welcome proposals using visual, graphic and photographic forms, including photo essays and original artwork which extend possibilities for the visual page. We are also interested in proposals for collaborations between artists and critics. *Performance Research* welcomes submissions in other languages and encourages work which challenges boundaries between disciplines and media. Further information on submissions and the work of the journal is available at: http://www.performance-research.net or by e-mail from: performance-research@aber.ac.uk.

Published 2008 by Routledge
2 Park Square, Milton Park, Abingdon, Oxon OX14 4RN
52 Vanderbilt Avenue, New York, NY 10017

First issued in hardbak 2018

Routledge is an imprint of the Taylor & Francis Group, an informa business

Typeset at The Design Stage, Cardiff Bay, Wales, UK and printed in the UK on acid-free paper by Bell & Bain, Glasgow.

Abstracting & Indexing services: *Performance Research* is currently noted in *Arts and Humanities Citation Index, Current Contents/Arts & Humanities* and *ISI Alerting Services*.

ISBN 13: 978-1-138-45887-1 (hbk)
ISBN 13: 978-0-415-49441-0 (pbk)
ISSN 1352-8165

Congregation

ISSUE EDITOR: CLAIRE MACDONALD

Performance Research 13(3), pp.1 & 188 © Taylor & Francis Ltd 2008
DOI: 10.1080/13528160902819554

'That is how we sorcerers operate. Not following a logical order, but following alogical consistencies or compatibilities. The reason is simple. It is because no one, not even God, can say in advance whether two borderlines will string together… No one can say where the line of flight will pass … Is it a new borderline? … Is it a way out?'[1] Manifesting archetypal forms in the landscape, symbolic representations of a complex of ideas, or a capture of non-visual forces as metalingual lines of flight. Points are nothing but inflections of lines – it's not beginnings and ends that count, but middles, twists in the path. Something 'new' could materialize at any point like a whirlwind twisting down a rocky path; yet this new thing exists within certain pre-determined parameters that limit the form produced: wings, thorns, horns, branches, strata, contours, twists, plaits, peaks, valleys, fluids, lichens, mosses, clouds, rocks …

[1] **Gillles Deleuze and Felix Guattari,** ***A Thousand Plateaus*** **(Athlone, 1988) p.250**

Editorial
Come Together

CLAIRE MACDONALD

We live in an age that is as vexed, fascinated and repelled by organized religion as it is at ease with personal approaches to the spiritual. As the philosopher John Gray wrote last year in the British newspaper The Guardian, 'An atmosphere of moral panic surrounds religion. Viewed not so long ago as a relic of superstition whose role in society was steadily declining, it is now demonised as the cause of many of the world's worst evils', noting that, conversely, the idea that 'the advance of science will drive religion to the margins of human life ... is now an article of faith rather than a theory based on evidence'.

A recent UK ad campaign on London buses, supported by a group of prominent secular writers and thinkers including Philip Pullman, Richard Dawkins and Polly Toynbee, declares that 'There's probably no God. Now stop worrying and enjoy your life.' When interviewed on BBC radio, Dawkins said that the word 'probably' had to be included for legal reasons. They would have preferred, as a definitive message, the assertion that there is no God, presumably thus enabling the public to breathe a sigh of relief, but implicitly assuming that the entire value and significance of religion rests on belief in God, and can be refuted in one fell swoop. The rest is not silence, but pleasure.

The conflict between faith and reason is deeply embedded within the histories of philosophy and politics, as well as within the arts and religion. The idea that religion can be refuted through the assertion of God's absence (or death) is a profoundly nineteenth-century trope, albeit a persistent one. Religion makes atheism in its own image. The novelist Philip Pullman has said that his own atheism was formed in the Anglican tradition and owes much to Milton and Blake, but while we choose to see belief, or faith, as the key - which raises core issues at stake in any discussion of religion and its place in, or its exclusion from, the social sphere - questions of practice and performance may be more useful and illuminating. While it is not only in the West that religion is sharply divided over questions of faith, the historian of religion Karen Armstrong has pointed out that, for most of the world's peoples, religion is less a matter of belief than a matter of practice. And as the performance scholar Kristin Dombek (whose work is on evangelical Christian practices) has said, looking at religious practices through the lens of performance is a way of sidestepping controversy and getting to the heart of what religious people do, as well as a means of understanding how spiritual and religious practices have informed contemporary art and artists.

An interest in practice rather than in belief is a key to the current issue of *Performance Research*. The emphasis here is on the sociality of religion - what it is and what it does as social performance, and how and where this intersects with and informs art-making. At the same time, I want to suggest that there is a gap in both practice and scholarship that might be further informed by allowing religion to speak to performance, questioning, analyzing and identifying in it in discursive and rhetorical

Performance Research 13(3), pp.2-4 © Taylor & Francis Ltd 2008
DOI: 10.1080/13528160902819232

terms structured through the praxis of organized religion.

Congregation takes its title from a question asked by Meredith Monk during a round table with Erik Ehn, Eleanor Heartney, Alison Knowles and Linda Montano on art and the spiritual, organized by Bonnie Marranca in New York in 2001. Both Monk and Marranca have long been interested in the ways in which spiritual practices have informed the practices of contemporary artists, and on this occasion Monk proposed the following as food for thought

> In 1996, the Union Theological Seminary asked me if I could make a service for the Guild of American Organists, Protestant organ players from all over the United States who were coming to New York for a conference. I enjoyed working on that very much because it was also a way of thinking about what exactly is a congregation? What is an audience? I was trying to do something that was between those two ideas.

Her question 'what exactly is a congregation?' intrigues me - firstly because she asked it (it is, after all, such a very unfashionable question). Secondly, because she begins to answer it through working on the question as a composer, an appropriate strategy for Monk herself as a Buddhist, a participating member of a non-theistic religion in which enlightenment emerges *through* practice. Thirdly, because it is in the slippery space between collective terms that identify us as participants in both secular and religious contexts that she finds the means to start.

In curating and editing an issue of *Performance Research* on religion - rather than on, say, the spiritual or ritual - I wanted to find a place to start that generated productive engagement among writers and artists with a variety of backgrounds and investments in the field. One of the things Performance Studies has continued to address is the position from which scholars and artists speak. Engaged, committed voices have been encouraged within the discourses of Performance Studies, as it has explored - and often unpicked - notions of authorial advantage, transparency and truth. Such an approach seems vital to a project such as this. In bringing together a range of voices, I have chosen to include work that originates within and from committed religious positions, as well as work by scholars and artists that engages with practices outside and beyond the limits of their own experiences.

What is at stake here is, I think, the importance of proposing a productive conceptual space in which to explore ideas and practices outside the often divisive contexts in which religion is discussed but also in which social and political decisions informed by religious beliefs and practices are made. Enlightenment notions of secularism have tended to suggest that secular decision-making space must be stripped of religious connotation - or religious investment - in order for reason to preside over superstition. Thinkers such as John Gray, John Milbank and others have begun to suggest that this concept no longer works. In the new world order we need to enable people to come to the table as fully engaged citizens, bringing jarring, dissonant, and passionate, voices from different belief systems. The challenge of such a congregational space is enormous, but this model also clearly makes connections between religious difference and the difference that has been proposed by sexual and identity politics. It offers a notion of the political and social spheres as rowdy, eventful and challenging, places in which radically different world-views can be present, and expressed. It is, clearly, the model that Barack Obama is working with, a model that has emerged in the two weeks or so that have elapsed since he became President of the United States.

This notion of 'congregation' - a gathering that is able to hold in dynamic exchange widely differing points of view and means of expression - has also informed this issue. *Congregation* is the largest issue we have yet published at *Performance Research*, reflecting the interest in the area, and I would like to thank the University of the Arts for supporting its publication.

We would have published more, but space did not permit. What I have chosen to include focuses on approaches to performance and to religion that suggest possible avenues for future research, that bring neglected areas of discourse into the field and that emphasize the contemporary. Historical essays have been included where they throw light on the contemporary in important ways and suggest new connections.

In an echo of an earlier period of intense engagement between religion, art and public discourse, the cover image is by the Catholic artist Corita Kent who, forty years ago, and with her sister nuns at the Immaculate Heart convent in California, created a radical, performative, art programme that brought, among others, John Cage, Buckminster Fuller and Jean Renoir to the convent. Her work has recently been celebrated and discussed in a book by Julie Ault for Four Corners Books, and I would like to thank the publishers for allowing me to use an image by an artist whose work has been described as 'some of the most striking - and joyful - art of the American 1960s'.

Congregation owes another, deeper, debt - to performance scholar Bonnie Marranca, whose interest in the field is evidenced in several issues of her own journal, PAJ, and in many interviews and articles. Her interviews, in particular with Peter Sellars and Meredith Monk, have informed my own thinking. Marranca and Monk have a new conversational interview in PAJ's most recent issue, PAJ 91, which contains a section on art and the spiritual. I like to think that *Congregation* complements this - and also that it acts to a certain extent as a corrective. In the rowdy, passionate space that exemplifies all engagements with religion, *Congregation* emphasizes the potency of collective, social performance, where *PAJ*'s *Art and the Spiritual* explores the personal and reflective. I look forward to taking part in the ensuing conversation.

I would like to thank the contributors to this issue, as well as thanking the writers whose work was read both by myself and by our peer reviewers but that we were not able to include. Inevitably, some of this writing has informed me, and I am grateful for the opportunity to engage with it. I also want to thank Sandra Laureri at CPR for her patient and thoughtful management of the editing and production process, and Richard Gough and Steve Allison for their editorial and design work.

Claire MacDonald
Andros, Greece, January 2009

ACKNOWLEDGEMENT

We would like to thank ICFAR and the University of the Arts London for their support of this expanded issue of Performance Research

International Centre for Fine Art Research

www.icfar.co.uk

University of the Arts London

www.arts.ac.uk

Body/Space/Worship
Performance theology and liturgical expressions of belief

PETER CIVETTA

THE PERFORMANCE THEOLOGY OF A GLASS OF WATER

The man preaching was new to the game. About to head to seminary, he had only preached a couple of times previously. As a litigator in his previous vocation, he felt comfortable speaking in front of people, but he struggled with the differences between preaching a sermon and trying a court case. In his law practice, lawyers spoke from podiums (not running around the courtroom as on television), so he wanted to leave the pulpit to, in part, make a break from his previous public speaking experience. His main concern remained content; he fixated on his sermon words as different from the words of litigation. He forgot, however, the impact of his body.

Standing on a platform one step above the congregation, he paced back and forth and began to speak. About a third of the way through the sermon, he walked over and picked up a glass of water. As he talked, he began to gesticulate with the glass, using it to emphasize his points. Sitting in the pews, I became more and more fixated on the glass. I kept waiting for how he would weave it into his sermon. I found myself leaning forward in anticipation; he had caught me with the glass. Then suddenly and without ceremony, he took a small sip and put the glass down. He was thirsty. For me, however, the entire sermon had been undermined. He had not taken into account the performance of the glass. He picked up a prop, used it within his delivery and then dismissed it as irrelevant. It provided something for a nervous hand. When we spoke afterwards, he said the glass had nothing to do with the sermon, neither in terms of content or delivery. It was just a glass, he said. However, theatrical reception doesn't work that way; the performer is not the only person who makes choices around the production of meaning. The moment that he picked up the glass and used it within the context of his speaking (as opposed to picking it up, taking a quick sip, and replacing it) became the moment that the glass entered the sermon. Three other people with whom I spoke after the service mentioned the glass (not at my instigation). They all said the same thing: 1) they expected the glass to have something to do with what he said, and 2) they missed a portion of the sermon when he put the glass back down as they reprocessed the fact that the glass was not, in fact, meaningful as they had expected.

I use this story as an example of the power of theatrical reception within liturgical/worship environments; how the same processes of reading performance exist in both. On stage and in a worship service, everything contains the possibility of being productive of meaning. I do not claim this idea as revelatory, yet little exploration exists of the relationship between performance and liturgy. I see this problem occurring for three reasons. One, liturgy connects to theology. To create effective worship in an Episcopal church (where this event took place), one focuses almost exclusively on textual resources, namely the *Book of Common Prayer* and various hymnals. The rubrics for the services outline standing, kneeling and sitting and,

Performance Research 13(3), pp.5-17 © Taylor & Francis Ltd 2008
DOI: 10.1080/13528160902819265

therefore, contain the perceived limits of performance. In other words, performance exists within a textual context. Follow the parameters of the book, and the service will be fine. Therefore, there appears a fundamental lack of awareness of the importance of the body, space and architecture except when demarcated within textuality. While certainly a generalization, this statement remains applicable in my work across a broad range of religious institutional experiences.[1]

Two, performance and theatre remain 'bad words' in many religious environments, raising resonances of falsehood and deception. Acting is about pretending to be something that you are not, they claim - the very opposite of how they view their own work. When I work with religious communities, I must scrub theatrical language from what I say, otherwise the words seem to create inherent stumbling blocks. When I work with clerics and speak of the importance of realizing the inherent meaning read from their bodies, the initial reaction is usually one of surprise. They put their stock in words, and they assume that the correct words can profess meaning no matter the context. They remain largely unaware of what messages their bodies' send. However, their congregants certainly don't.

Third, the predominant way that performance scholars have approached liturgy is to theatricalize it, to treat it as a performance in line with other cultural performances. However, this approach creates a problem because of its status as worship. In my experience, to use a theatrical viewpoint loses the importance of worship as a foundational aspect of the performance. Just as religion's ignoring of performance has inhibited its ability to fully understand itself, an over-reliance on performance misses this key dynamic. These people attend for worship, not for a show (although they certainly can and do overlap).

In my work, I have needed to find a new way to approach these types of liturgical performances and communities, and I use a term that I call *performance theology*. Performance theology explores the relationship between stated belief and lived experience. It is the study of what people say they believe, how they choose to live their lives, and what role performance plays in that transaction. Performance theology also provides a methodology, a means of engaging people about their beliefs without (pre)judging those beliefs. For example, I know an intelligent and accomplished woman who sees angels. Apparently, there are many different forms of angles beyond the swaddling babes with a bow and Matt Damon and Ben Affleck in *Dogma*. She very calmly and articulately describes angels she sees, as you might describe a table or chair.

Normally, we find few options on how to respond - she is either crazy or, well, she's crazy. I guess there's only one option, and it was my initial, unchecked reaction. Then I thought: what if she does see angels? How does that change how I deal with her? Instead of approaching her from my perch atop intellectual condescension, what if I accept that it is true *for her*?

First, it doesn't mean that I see angels or that I even have any sort of opinion on their material existence. It has nothing to do with what I do or do not believe at all. It becomes clearly demarcated as her belief. Second, if I accept the fact that physical angels appear for her, then I am free to explore how she came to this realization and how it impacts on her life. If I don't, if I continue to judge her belief as untrue or crazy, then I have limited, if not eradicated, the capacity to engage this person. Performance theology does not validate a person's belief structure; it honours it as true for them as a means of investigating what impact that belief holds. Therefore, performance theology allows me to explore the efficacy and agency found in liturgy through performance while keeping ever-present the role that worship plays for those people involved. Using performance theology as a lens, this article explores the experience of meaning-making for congregations within religious performances. I will utilize two case studies. The first explores the role that the body plays in establishing religious beliefs, using Muslim prayer as my

[1] I have worked ethnographically in religious institutions across faith traditions and have led workshops and conferences on liturgy and preaching within religious communities.

touchstone. The second looks at the profound impact that space and architecture have over liturgical experiences at an Episcopal Church. In sum, my aim lies in establishing a widening exploration of religious experience for performance scholars, including the body, space and worship as foundational elements.

THE DANCE OF PRAYER

Imam Kasim Kopuz serves the Al-Nur Mosque in Johnson City, New York, where every Friday afternoon he leads his congregation - exceptional for its ethnic and national diversity - in Jum'ah Prayer. I spent over three years at Al-Nur (2001-4), and I learned that upwards of fifty different countries are represented in the mosque. The Imam could rattle off twenty to twenty-five different nationalities in less than a minute's time, including all of the Arab world and most of Africa and South Asia. 'The whole Muslim world is present here,' he would say. This level of diversity remains quite rare, even in cosmopolitan cities where separate mosques get established for different ethnic and regional groups. One city may have an Arab mosque, a Turkish mosque and an Indian subcontinent mosque. While people certainly may pray at any of them, they often have a difficult time getting their voice heard within the administration of the mosque. Even 'international' mosques rarely possess the diversity of Al-Nur without some degree of tension or conflict coming from that diversity.

Of those people who regularly gather for Jum'ah Prayer, approximately thirty percent are of African or African American descent. The majority of the mosque members have Arab and South Asian backgrounds, with less than five percent Caucasian. The largest single constituency comes from the Indian subcontinent, along with many Kurds. I would estimate that, on average, approximately thirty to thirty-five percent of all attendees wear traditional Islamic robes, although fewer also wear turbans. The diversity of clothing reflects the diverse economic backgrounds of members of the community.

Since Jum'ah Prayer takes place during a normal workday, many people come directly from their jobs, to which they return after services. Therefore, the community often includes men in nicely tailored suits, hospital employees in surgical scrubs, industrial workers in heavy overalls, and a large number of university students.[2]

Jum'ah Prayer begins with a *khutbah* (sermon). The start time for the khutbah is 1:10pm, although the first azan (call to prayer) occurs generally around 1:00pm. The azan, performed before all five daily prayers for Muslims as well as before Jum'ah Prayer, consists of seven statements. 'Allah is most great. I testify that there is no god besides Allah. I testify that Mohammed (s.'a.w.)[3] is the messenger of God. Come to prayer. Come to salvation. Allah is most great' (Antoun 1989: 79). More than a perfunctory reminder, the azan, according to Kenneth Craig, 'has to do with present achieving as well as given status. Muslims are challenged, so to speak, to become what they are' (Craig 2000: xi). The azan serves as a call to re-embody their faith, to live it back into their bodies, to move it from thought to action. When responding to the call to prayer, 'Muslims have actualized their Islam in their response' (Craig 2000: 96).

After the first azan, most of those people in attendance rise and perform their personal salat (prayer). Many variations exist in the performance of salat, including the types of prayers spoken and the number of times each section gets performed. Outside of the communal performance of Jum'ah, salat is performed silently. Therefore, I can offer here only a general idea of possible prayers as demonstrated by this community and explained by the Imam, without making any claims of its normative practice within[4] Islam.

Salat begins by first orienting your body towards the *Ka'bah*, the sacred mosque in Mecca, which is a strict requirement for the prayer to be considered valid. The stripes on the rugs in Al-Nur indicate the correct direction, facing an open archway called the mihrab (niche, /recess).

2 My experience rendered here is primarily based upon contact with male members of the congregation. This mosque uses separate worship spaces and entrances and exits for the sexes, and I had very little contact with women during my time there. Therefore, the ideas expressed here may only be extended to the male population of the mosque.

3 This abbreviation means 'May the blessings and peace of Allah be upon Him'. It is used every time the name of the Prophet Mohammed is mentioned or written by a Muslim.

4 Outside of the khutbah, the entire service at Al- Nur is in Arabic. The uncited quotations are from translations made for me by Imam Kopuz.

'The recess serves as a focal point that sets the faces of the faithful on the line to Mecca where they find spiritual rendezvous with the rest of their fellow Muslims. By the mihrab the congregation is consciously set on one of the radii from the gravitational centre of Islam' (Craig 2000: 114).

The worshippers raise their hands to both ears, palms facing outwards, while they state the greatness of Allah and their *niyyah* (intentions) for this prayer. After this action, all worldly 'talk or movement is forbidden' (Saqib 1986: 32). 'The need for the "intention" makes evident that true prayer cannot be perfunctory or mechanical. It is a means of defense against inattentive and external performance. The movements no doubt become habitual, but habituation should not lead to forgetfulness' (Craig 2000: 99). The worshippers drop their hands, either to their sides or clasped at their navel, offering a prayer of supplication to Allah. At this point, the opening *Surah* (verse) of the Qur'an is recited.

The next step includes a *raku* or bowing at the waist, while proclaiming, 'Glory be to my Lord Who is the very Greatest.'. They may repeat this proclamation, but, according to a *Hadith* (sayings of the Prophet Mohammed), it must be done an odd number of times. When the worshippers stand up, they state, 'Verily Allah listens to one who praises Him,', while raising their hands to their ears again. They may include additional prayers here and repeat as often as the worshipper desires (but always an odd number). The next action, the *sajdah*, consists of a full prostration, where seven parts of the body must touch the ground: 'The forehead along with the tip of the nose, both hands, both knees, [and] the bottom surface of the toes of both feet' (Saqib 1986: 43). In this position, the worshippers say, 'Oh Allah, glory be to You, the Most High.' Other prayers may also be added. Upon completion, the worshippers should raise their heads, assume a seated position, and offer a prayer asking Allah for forgiveness. Another sajdah follows, mirroring the first, before they resume the seated position. They may add a number of prayers and Surah at this point, before the worshipper states, 'Peace be on you and the Mercy of Allah', while turning their face first to the left and then to the right. The series thus described, called a *rakkat*, may be repeated any number of times. During Jum'ah Prayer, the community together performs the sequence twice.

On two successive khutbahs in June 2003, Imam Kopuz extolled the virtues and rewards of communal prayer. Although performed individually at most times during the day, salat carries crucial communal weight. Salat prepares the believer to deal with the world, to live out the path prescribed by Allah. The purpose of salat, the Imam said, lies in connecting worshippers to Allah, so that believers can carry that connection out into the world, i.e., they bring Allah into the world through salat. Salat supports and guides through discipline and devotion to Allah, creating a foundation for all social conduct.

After doing their personal salat, a second azan is made, and at its conclusion, the Imam steps to the minbar (pulpit) to deliver the khutbah. Once completed, the Imam descends from the minbar, and a short azan serves as a transition into Jum'ah Prayer. At this stage, members of the congregation, spread throughout the room, arise and form lines. They take their positions, using the striped carpet to orient themselves, with the Imam reminding them to stand close together, shoulder-to-shoulder. The Imam then leads their communal Jum'ah Prayer.

It remains striking to behold this diverse community perform this sense of unity that they have found through their beliefs. Every week they display the ummah (unified Muslim community) through performatively standing close together, touching each other, in prayer. On many occasions, I witnessed people inviting others to fill gaps in the lines. Even teenagers and young children take the performative event seriously.

Another important element of this experience concerns the Islamic concepts of innovation and *bid'ah* (bad, not confirmed by Islamic law). In a prayer before beginning his khutbah, Imam Kopuz states:

> Verily, the best of the words are those of Allah, and the best of the guidance is the guidance of Prophet Mohammed, peace and blessing be upon him. The worst of the matters are those innovated, and every innovated thing is *bid'ah.* Every bid'ah is an act of misguidance, and every such misguidance deserves the Fire.

I previously knew of the negative connotations of innovation in Islam, but I remained surprised that Kopuz chose to give it such a prominent placement, immediately before he begins every khutbah. At our next meeting, I asked him about the concept and why he finds it so important. He said many non-Muslims do not correctly understand this concept of innovation; they assume that it means that nothing in Islam can ever change. In fact, he said innovation holds a very narrow meaning for Muslims, concerning only ritual worship. The actions of ritual worship in Islam remain forever a mystery. 'We can never truly understand why we do certain things. What is the point of standing here, bending there, and prostrating ourselves? We can never know why Allah has us pray five times a day.' Since these things have come from Allah, they may never get altered in any way because to do so would be to place yourself ahead of Allah. He gave the example of giving money to charity. When performing that act, we understand what we do and its worthiness. Yet, no such knowledge is possible with ritual. He said people may realize spiritual insights or understanding while performing rituals, but that effect may come from other experiences as well. Ritual doesn't contain the only link to that experience. Without divine understanding, humans cannot alter ritual worship prescriptions in any way. To do so would superimpose the will of humans over the will of Allah. However, this Qur'anic directive does not effect actions outside of ritual. For example, cars and the internet present no problems for Muslims. Therefore, using a microphone for a khutbah, since not part of the ritual prescriptions, does not get considered bid'ah.

This concept also explains why I could find nothing of substance ascribing meaning to the physical actions of salat. In a review of many prayer books that outline the practice, every one simply states what must be done without commenting as to why or what meaning it holds. It simply *is*. This pattern also confirms what so many of the people that I met with state, or rather wouldn't state. No one at Al- Nur would offer a definitive meaning behind any of the actions of salat. The first time I witnessed Jum'ah Prayers, I wondered about what it meant to hold their hands to their ears, but no one would ever offer a significant reply to this question. I found a few interpretations, but they all came from outsiders to the faith. For example, Craig explained prostration in the following terms: 'Prostration, in particular, proclaims and serves to actualize a totality of surrender. The face, the proudest thing in human personality, comes into contact with the dust, the lowest thing in nature. The physical thus embodies and expresses the spiritual' (Craig 2000: 98). Since Islam means submission, it seems cohesive and clear to portray the prostrations of salat as showing submission as Craig does. However, while no one I spoke with disputed that reading, it remains, almost, beside the point. Instead of viewing it as simply an act of submission, this community remains open to the fact that they can never fully know what it means, thereby increasing the possibility of other experiences. These notions of innovation and bid'ah require that Muslims remain open to the experience of salat instead of merely commemorating specific described or describable meanings of it. Without recourse to definitive judgements as what the movements represent, they must sit and experience them for what they are. It becomes a lived action of belief, a performance theology.

Still, as an outsider, I searched for ways of understanding what the movements can mean, and I found a connection in an unexpected source - famed dancer and choreographer Martha Graham. Salat helps Muslims to connect deeply with Allah, which in many ways means to connect with the totality of what Allah represents. The practice, instituted by the

Prophet Mohammed, has remained constant ever since that time. Salat carries the weight of the past as well as being a means of knowing Allah in the present. Martha Graham talks about finding meaning in the movements of dance in the following terms:

> The only thing we have is the now. You begin from the now, what you know, and move into the old, ancient ones that you did not know but which you find as you go along. I think you only find the past from yourself, from what you're experiencing now, what enters your life at the present moment. We don't know about the past, except as we discover it. (Graham 1998: 70).

Taken with salat, I take this idea to mean that the access to Mohammed and Allah comes through the present moment and through the bodily expressions of belief. The movements done in the present moment have the ability to connect the believer back to the past, but this action happens on an individualized basis. In this way, salat empowers on a personal level, as meaning comes through each believer's experience of the past through the present. Each person does not try to mimic a particular reading or understanding of the movement; Islam goes out of its way to not create such a doctrine. Instead, there is trust that the experience of the body will lead the believers towards the Holy. They do not need to prescribe meaning; it remains theirs to work out and experience on their own. Graham states, 'You get to the point where your body is something else, and it takes on a world of cultures from the past, an idea that is very hard to express in words' (1998: 70). The proscriptions of bid'ah and innovation insure that 1) the movements won't change and 2) meaning for them will not come from others but must be achieved individually.

In what way does this experience happen? Due to the heavy proscriptions of the physical actions included in the prayer, learning the movements of salat will take a while. It takes even longer before the movements become natural and decided, the state where this community told me prayer can occur. Graham talks about the training of a dancer, a process she says takes about ten years, but it is interesting to read the learning of salat across it. 'First comes the study and practice of the craft which is the school where you are working in order to strengthen the muscular structure of the body. The body is shaped, disciplined, honoured and, in time, trusted. The movement becomes clean, precise, eloquent, truthful. Movement never lies. It is a barometer telling the state of the soul's weather' (Graham 1998: 66). In these terms, it takes time for the salat movement to find ownership in the body, to become a part of it. Training leads to trust that leads to truth, and, in the end, that trust allows for (or creates) the experience of connecting with the Holy. Movement never lies, and these movements, in Islamic theology, will lead you to Allah. The prayer becomes the meaning; the movements become the experience of God. This process requires learning and engagement, and it also requires trust and release. Graham states: 'I believe that we learn by practice. Whether it means to learn to dance by practicing dancing or to learn to live by practicing living, the principles are the same' (Graham 1998: 66). When mapped onto Muslim prayer, it reads: *whether it means to learn to pray by practicing praying or to learn to believe by practicing believing.* Graham continues: 'In each it is the performance of a dedicated precise set of acts, physical and intellectual, from which comes shape of achievement, a sense of one's being, a satisfaction of spirit. One becomes in some area an athlete of God' (Graham 1998: 66). Hence, a true believer.

From this experience, I gain increased recognition of belief as not wholly thought, not a solely conscious and intellectualized process of discernment and acceptance (or rejection). Performance theology lives as a bodily function; how these people choose to live their lives is in part dictated by what they learn from their bodies (not their minds) in the act of prayer. Belief does not exist merely as a set of dogmas and presuppositions; it is a lived experience. Belief not only gets expressed by the body - an after-effect of previously determined ideas - but comes from the body as well. Salat at the Al-Nur Mosque contains a bodily expression of belief, and the worshippers' role as worshippers means to engage and embody that expression if they want to get in touch with that belief. This link of prayer offers them unity as a community, with such enormous diversity, through their common bodily actions, but it also allows for and fosters their sense of individuality, as no one, not even the Imam, may dictate what the movements may mean for them. The physical act and actions of prayer both unites them and provides for their uniqueness. Thus, the body becomes a primary site of performance theology, and it provides an opportunity for further examinations into the relationship between bodily religious practices and lived belief.

THE FONT GUILLOTINE

Episcopal liturgy explores notions of space, but within very precise renderings. The directives (rubrics) of liturgy tell the clerics and laity what to do with their bodies when: stand here, raise arms there, process from A to B, etc. Each of those movements and locations becomes bound with symbolic meaning: the altar as representation of the table of the Last Supper. Even sides of the church have traditional symbolic meaning. However, the combination of physical space and conceptual space remains notably absent from most liturgical thought, and it made me wonder about the role of worship space. Can religious space dictate the experience of the people worshipping? I do not mean merely sight lines, but whether in fact the space creates not only the worship but the congregation itself.

I recently led a workshop at Grace Episcopal Church in a suburb west of Chicago (2007). On a break, a group of us entered the sanctuary, both a beautiful and imposing space. I returned to the church a few weeks later to photograph the sanctuary and to speak with the current rector of the church and a major member of the Diocesan staff who worships and serves there regularly. Our discussions and my experience provided me with the material to approach my questions about religious space.

In *Thirdspace*, Edward Soja details and adds onto Henri Lefebrve's work on spatiality. Both seek to add space to the binary of historicality and sociality, not eliminating or marginalizing those important modes of discourse, but rather adding an important missing element. As Soja states,

> [A]ll social relations become real and concrete, a part of our lived social existence, only when they are spatially 'inscribed' - that is, *concretely represented* - in the social production of social space. Social reality is not just coincidentally spatial, existing 'in' space, it is presuppositionally and ontologically spatial. *There is no unspatialized social reality*. (Soja 1996: 46, emphasis in original)

To fully understand how social reality works, we must pay closer attention to the role of spatiality. For example, while liturgical rubrics outline movements and locations, no accommodation gets made for differences in physical space. In other words, the rubrics create a one-size-fits-all mentality even though church architecture can run the gamut between high cathedrals and converted commercial office spaces. Liturgics doesn't ignore spatiality, but it marginalizes space into two of the three notions outlined by Lefebvre: perceived and conceived. Lefebvre's lived space - and Soja's Thirdspace - remains under-explored in terms of the role that space plays upon worship.

Firstspace, developed from Lefebvre's perceived space, focuses on the materiality or physicality of space, its dimensions and nameable attributes. Taken to its extreme, this perception of space appears in the proliferation of GPS navigation systems. Space gets mapped out and thus becomes knowable. Space appears as a material object that is describable. This perspective provides one of the dominant forms of approaching religious space, where physical space becomes the manifestation of understanding. Grace Church presents a perfect example of this dynamic.

They broke ground on the 11,000 square- foot brick and Indiana limestone nave in 1898 and dedicated the sanctuary in 1905. The design came from noted architect John Sutcliffe (they apparently passed over famed local Frank Lloyd Wright due his scandalous personal life). Figure 1 shows the interior of much of the space.

• *(left)* **Figure 1: Grace Episcopal Church, Oak Park, Illinois**

• *(right)* **Figure 2: Clerical Areas Of Worship Space**

If it seems familiar to you, the sanctuary has appeared in such films as *Home Alone* and Robert Altman's *A Wedding*. My first impression fixated on the largeness and openness of the space. Although not comparably wide, the length and height of the church create the perception of a vast and imposing place (note that despite the long angle photograph, no ceiling is visible). At the time of its design and construction, the Episcopal Diocese of Chicago did not have a cathedral, so this space reflects a specific plan and desire (unsuccessful) to achieve that vaunted designation. In a space such as this one, distance becomes the operative ingredient. One can barely discern the high altar both due to its distance and because of the elaborate, wooden, rood-screen-ish structure that separates it from the congregation. Given the Eucharistically-centered tradition of the Episcopal Church, this distance and screening are quite notable. In most Protestant churches, a pulpit stands front and centre, but in Episcopal churches (as in Catholic ones), the altar maintains this prominent placement, with the pulpit off to the side.

Figure 2 shows a closer vantage point of the clerical areas of the worship space.

From this perspective, a new distance becomes apparent. There is a distance of about 20 feet between the first pew and the chancel steps leading to the pulpit (on the left of the picture), the lectern (on the right where readings and announcements are given) and the back altar and choir area. Therefore, this picture captures the closest vantage point to the Eucharistic celebration at the core of their faith, a distance of 40-50 feet. Centered in this picture, one sees a covered table that also serves at an altar. However, they only use it one Sunday a month; the rest of the Eucharistic services occur at the high altar at the back.

The entire space creates an aura of the sacred, seeking to capture in smaller form the majesty of Europe's cathedrals. In short, the design of this spectacular space desires to connect one to the grandeur and majesty of God. This space embodies both an epic God and a distant God, making the individual feel small and somewhat insignificant. Therefore, this space sends out very powerful messages about the nature of God and of the congregation's possible relationship to this God. Does this space allow for intimate connection with God? God as authoritative Father is certainly present, but what of the notion of *Abba*, loosely translated more as Daddy? The physical space at Grace Church circumscribes and dictates the range of potential experiences of the Holy found within it. Other encounters remain possible, but to achieve them, one must battle the very architecture. For example, the current Rector of Grace is the Reverend Shawn

Schreiner. Although she is a petite woman, the space in no way intimidates her. Her distinguished career leaves her comfortable in most (if not all) worship environments. Deeply learned in liturgy, she fully understands the dynamics of space in discerning possible choices for the performance of liturgy. However, these facts don't eliminate the problems that she faces here.

The pulpit, seen in Figure 3, fits the space with its impressive style.

• *(left)* **Figure 3: Grace's Pulpit**

Unusually large and elaborate, it certainly carries the weight of authority. Roxanne Mountford, in her book *The Gendered Pulpit*, explores the rhetoric of architecture for preaching. She argues that pulpits carry physical expectations of a male presence, with their architecture quite literally designed for male bodies. The size of the walls and level of the lectern get proportionally dictated by the size of men. Women often find themselves dwarfed in them. Mountford cites Lefebvre talking about city landscapes, writing: 'Verticality and great height have ever been the spatial expression of potentially violent power' (Mountford 2003: 24). We can see pulpits, such as Grace's, as expressions of clerical authority and power. In the vernacular, pulpits present large, erect structures culturally and institutionally associated with male authority and, um, power. For a sermon to access its traditional power, it requires its traditional placement: the pulpit.

This pulpit provides the preacher with authority, raising the person, literally, above the congregation and surrounding that person with ornamentations well beyond those found in places where the congregation sits. Canon Randall Warren of the Episcopal Diocese of Chicago has served often at Grace Church. The side-by-side Figures 4 and 5 show the dramatic difference in what the space of the pulpit conveys. In short, Reverend Warren 'fits' in the space in a way that Reverend Schreiner does not.

• *(right)* **Figure 4: Reverend Warren In Pulpit**

• *(right)* **Figure 5: Reverend Schreiner In Pulpit**

The pulpit arrays around Warren, while it almost entirely consumes Schreiner. Warren gets to maintain full expressive use of his arms and hands, yet Schreiner's remain trapped behind the wall of ornamentality. A closer view (Figure 6) shows just how completely her body gets absorbed by the male-dominated space. Here the light fixture appears to separate her head from her body, and the lectern eclipses even more of her body's shape and expressiveness.

Leaving the pulpit doesn't simplify the

• *(light)* **Figure 6: Reverend Schreiner In Pulpit, Close-Up**

problem. Her height makes preaching from the centre aisle pragmatically difficult, and preaching raised up on the chancel steps doesn't work well either. While this does improve sight lines and allow for more bodily movement than the pulpit, a single person (even Warren, as shown in figure 7) gets lost within the majesty of the space. No authority there.

• *(left)* **Figure 7: Reverend Warren On Chancel Steps**

Also, even if the preacher leaves the pulpit, the pulpit doesn't leave the space or the consciousness of the worshipping community. In this picture, the pulpit looms, continuing to project the spatial implication of the proper place to preach. Even when silenced, the pulpit speaks.

Therefore from the perspective of Firstspace, or perceived space, Grace Church creates a specific (and gendered) worship environment. While it sends a message of sacredness, this specific type of sacredness predetermines some of the worship practices that can take place within it. Its glories as English Gothic Revival architecture also contain its limitations and even its shadow side for the congregation present.

Soja's Secondspace deals with Lefebvre's conceived space, stating that

> spatial knowledge is primarily produced through discursively devised representations of space, through the spatial workings of the mind. In its purest form, Secondspace is entirely ideational, made up of projections into the empirical world from conceived or imagined geographies.
>
> (Soja 1996: 79).

Secondspace can be likened to semioticians making space symbolic, 'a world of rationally interpretable signification' (1996: 79). The connection to religious space remains clear because it directly links space to God. It offers the theologization of sacred space, where everything within the space (as well as the space itself) becomes interpretable with religious meaning. In this way, Firstspace and Secondspace work hand-in-hand in religious spatiality. However, while inherently conceptual spaces, this theology of space can prove as restricting as the physical space.

At the back (or rather the entrance) of Grace Church, the baptismal font rests in a relatively small alcove directly behind the rows of pews. The elaborate marble font stands on two marble steps and has an ornate carved wooden cover made to resemble a church spire (see Figure 8).

• *(right)* **Figure 8: Grace's Baptismal Font**

A beautiful work of art, the font fits perfectly within the conceived space of the church at the time of its building. At the time of construction, baptisms were private events, held outside of normal Sunday liturgies. Therefore, on a conceptual, Secondspace level, the font exists merely to remind the congregants as they pass by during regular services of their own baptism. However, the theology of baptism has changed radically since then, and now normative practice holds all baptisms within the context of a Sunday liturgy. In fact, both Schreiner and Warren profess to holding private baptisms only in extraordinary circumstances.

So how has the space adapted to this change? Not so well. First of all, to see the service, the congregation must completely turn around, leaning into their own pews. Second, the alcove comfortably holds a single family preparing to baptize a single person. Now multiple persons receive baptism at one time in what becomes part of a communal celebration where the congregation also renews their own baptismal covenant. Positioning the multitude of people remains challenging as does creating clear ways for the congregation to view and participate. Third (and most humorously) comes from the font itself. Again, originally the family would stand around the cleric (against the back wall) during the baptism. Now the rear area becomes more of a 'stage', the only vantage point from which to show the baptism to the people. In this instance, a short woman actually has an advantage, as Figure 9 shows Warren in the normal position that he would take to begin a baptism.

• Figure 9: Reverend Warren Behind Baptismal Font

Yes, his head gets cut off, an effect even more disconcerting when viewed from farther back where the congregation sits. You just see arms and a torso flailing about while a disembodied voice intones about the most significant of sacraments. Reverend Schreiner fits more easily behind the font, but she struggles to lift the infant, who often is small child, high enough to perform the baptism. In this way, conceived space, although part and parcel of religious space itself, can also become complicated and difficult when placed within the parameters of Thirdspace / lived space. However, in Thirdspace / lived space, I find the more fascinating and relevant connections between space and congregation than merely logistical concerns of finding a place from which to preach or baptize.

Soja writes:

> *Everything* comes together in Thirdspace: subjectivity and objectivity, the abstract and the concrete, the real and the imagined, the knowable and the unimaginable, the repetitive and the differential, structure and agency, mind and body, consciousness and unconsciousness, the disciplined and the transdisciplinary, everyday life and unending history. (1996: 56-7).

Thirdspace meshes the physical space of Grace Church and the theology of the Episcopal Church within the confines of liturgical performance. None of these three happen independently of the others. Liturgy only happens in a particular space within the framework of a conceived theology. Therefore, the lived experience of the congregation appears (at least partially) to be dictated by the space in which they worship. Thirdspace concerns do find some articulation within the seminal work on Episcopal liturgy by Charles Price and Louis Weil.

> When we stand in the presence of God, we not only engage in the moral and intellectual activity of assigning values. More immediately and dramatically, we do something with our bodies. Ritual action, as well as decision making, is an inalienable part of worship. Worship is an activity of human beings in their complete selfhood, flesh in inextricable unity with the spirit. The holy God demands of us a total response. (Price and Weil 1979: 19).

Worship partakes of lived or Thirdspace because it encompasses physical and conceptual space as well as the living, breathing bodies and lives of the people partaking in it.

Schreiner describes this congregation as focusing on visual aspects of worship over liturgics. For example, they prefer her to use the back altar instead of using the table altar that would allow them much closer proximity to the communal experience at the core of their beliefs. They provide two main reasons. First, they designate the back altar as the place where 'it is supposed to take place'. Much like the looming pulpit, the high altar doesn't disappear when they use the closer one, and only the cleric (who faces the congregation) doesn't have the constant visual reminder of that 'correct' space. The pulpit and high altar scream out for use, incredibly conspicuous even when avoided. Thus, their initial objection is Firstspace, the perceived and 'natural' use of the space.

Second, Schreiner expresses that people cite the choir as a main reason for wanting to use the back altar. On its face, that choice appears curious, since surely they can hear the choir clearly from all places within the church. However, if you return to Figure 2, you will see the specific configuration of the back altar space. The choir pews face each other in Anglican chant style, so when a congregant comes to the altar rail at the back altar to receive communion, the person literally walks through and in the midst of the choir. They cite the feeling of walking through the music as incredibly important to their experience of Eucharist. Therefore, the Secondspace, or conceived expression of the space, also dictates how they understand spatiality and their experience of worship.

This final point shows the workings of the space upon the lives of these congregants. Schreiner explains the theology of the congregation as 'internal over external.'. 'They come to church to be fed'; church exists as an individual experience that tends to ignore the pragmatics of doing God's work in the world. She describes it as a dependent notion of faith - it is for me, so I may be fed. The message of 'being fed to then see to the feeding of others', a fuller embrace of mission and outreach to the broader community, has not been easy for this community.

It remains important to note the incongruity of this perspective within the theology of Episcopal liturgy and this particularly community. Price and Weil emphasize the critical importance of community in the creation and experiencing of worship.

> The response of faith to the actions of God on our behalf must not, and indeed for Christians *cannot*, be conceived in narrowly individualistic terms. The response of faith is a personal response within the community of faith, the church.
>
> (Price and Weil 1979: 30, emphasis in original)

They specifically warn against this very type of individually-centered worship. 'The problem is the separation of liturgy from the life of the Christian person and from the life of the Christian community. Liturgy then becomes objectified and depersonalized, a *printed text*, rather than the expression in action of a community of faith' (Price and Weil 1979: 31). Within this argument, it is not merely a printed text, but also a physical space that predetermines and separates the people from the experience of their faith within the confines of Episcopal practice.

Also, the church rests in one of the most liberal and politically active suburbs in the area. In other words, this community remains engaged and pro-active in almost every aspect of their lives (school systems, local governance, eco-friendly living, etc.) with the notable exception of their lives as a faith congregation. Why? The space doesn't allow it.

The distance required by the space creates a sense of invisibility; you get lost in the hugeness of God. The anonymity and seclusion dictated by the space for the congregation puts the emphasis for worship on individuality and visuality. Their space expresses God as far away and masked (the high altar with the wooden rood screen blocking), a space not easily conducive to communally-based experiences. Therefore, they have embraced a theology of individuality. Everything remains far away and formal, creating a formal, individualized notion of faith.

Price and Weil state, 'Worship always has two phases: in one phase, the community gathers to celebrate its liturgy in the usual sense of the word, its praise and thanksgiving, its penitence and adoration in the presence of the crucified and risen Lord' (1979: 56). This passage accurately depicts the worship and congregation at Grace. Beautiful church services take place here with many people of deep and abiding faith. Their space - both perceived and conceived - fosters, nurtures and sustains this type of worship. However, Price and Weil continue: 'in the other phase, the community lives in the world according to the love and power which come from him, in obedience and trust' (1979: 56). Here the space in which Grace worships problematizes their own experience of faith, as it overpowers communal instincts with its grandeur. Intimacy - and its consequent outreach - remains difficult to manifest at Grace.

In this way, spatiality at Grace Church possesses its own performance theology, and that performance theology has determined to a large extent the possible performance theology of the worshipping congregation. Not only people express a performative interaction between God and humanity: space does as well. In the case of Grace Church, the performance theology of the space has predetermined the horizon of expectations for the congregation, to borrow from Hans Robert Jauss. To gain their own control, they will probably need to embrace their own performative potential and challenge that of the space; otherwise, the church itself will continue to exert more control than even its clerics or doctrines. (See Figure 10.)

• **Figure 10**

REFERENCES

Antoun, Richard T. (1989) *Muslim Preacher in the Modern World: A Jordanian Case Study in Comparative Perspective*, Princeton: Princeton University Press

Craig, Kenneth (2000) *The Call of the Minaret*, 3rd edn, Oxford: Oneworld Publications.

Graham, Martha (1998) 'I Am a Dancer', in Alexandra Carter (ed.) *The Routledge Dance Studies Reader*, London: Routledge.

Mountford, Roxanne (2003) *The Gendered Pulpit: Preaching in American Protestant Spaces*, Carbondale, Illinois: Southern Illinois University Press.

Price, Charles P. and Weil, Louis (1979) *Liturgy for Living*. New York: Seabury Press.

Saqib, M. A. K. (1986) *A Guide to Prayer in Islam*. 2nd edn. London: Ta-Ha Publishers.

Soja, Edward S. (1996) *Thirdspace: Journeys to Los Angeles and Other Real-and-Imagined Places*, Oxford: Blackwell Publishing.

Spectres of Exchange
Rights and resources in *loisaida* liberation-theology passion play performance

LARA D. NIELSEN

In 1998 a coalition of three primarily Spanish-speaking Lower East Side ('Loisaida') Roman Catholic communities of Puerto Rican descent - locally, 'Nuyorican' - publicly performed the passion play, a ritual and dramatic reenactment of what is referred to as the Passion of Christ, in the raw March streets of New York City. The collaborating churches were María Auxiliadora, San Emérico and Santa Brígida. A local parish priest announced that on this day, Good Friday, 'our neighbourhood becomes our church'. The passion play performance was practiced through a bilingual call and response, choreographing the movements and voices of various kinds of participants in the action, from casual observers to costumed 'en/actors', with those of the priests. Through performance, the fourteen Stations of the Cross that constitute the iconic 'chapters' of the passion play mapped the neighbourhood, beginning with the first station (where Jesus is condemned to death) outside the church of María Auxiliadora on 12th Street between Avenue A and First Avenue. The response to the passion play's narrative of the first station, scripted in a pamphlet distributed free to anyone on the street who wanted it, reads as follows:

Gentle Jesus, we are outraged that you are condemned to die
with no one to defend you.
As we begin this journey with you to Calvary,
we ask for the deep faith necessary to see you
in the homeless, the hungry and those without health care.
We want to walk with them as we walk with you.

From there the costumed procession, with its crowd of parishioners and interloping publics, moved a few blocks east to reenact the second station (where Jesus receives the cross) and the third station (where Jesus falls for the first time) to pause outside the church of San Emérico, at 12th Street between Avenues C and D, for the fourth station (where Christ meets his sorrowful mother). This passion play performance articulates repertoires of resistance through a ritual remembering and recall of what Michel Foucault has called, after Marx, the materiality of political 'spirit'. Foucault was fascinated with the possibility for progressive uses of religion to oppose the perversions of state power: 'religion's role,' he imagined, 'was to open the curtain' for the next, and ostensibly better history.[1] This narrative amounts to a rational 'rejection of a modernization that is archaism'; he was looking for organized resistance to globalizing totalitarianisms. For my purposes, it is enough to notice that his formulation is rooted in a simple and yet crucial desire: to revisit the biopolitics of profane/sacred exchange economies, as the Loisaida passion play performance so cannily suggests.[2]

As I am especially interested in the ways that they depart from traditional passion play liturgies, it will be important to cite several scripted responses to the stations of the cross in detail, which alternate between English and Spanish for the Loisaida community. For the first fall of the Viacruces (way of the cross), the third station response likens everyday indifference

1 In his coverage of the Iranian Revolution in 1978, Foucault reminded readers of the Italian newspaper *Corriere della Sera*, 'People always quote Marx and the opium of the people. The sentence that immediately preceded that statement and which is never quoted says that religion is the spirit of a world without spirit'. In Marx, the quote reads: 'Religion is the sigh of the oppressed creature, the heart of a heartless world, just as it is the spirit of spiritless conditions. It is the opium of the people' (Marx and Engels 1975: 175). This is what Foucault saw in 'a powder keg called Islam'. Unsurprisingly, he was soundly defeated in public arguments (Rodinson in Afary and Anderson 2005).

2 From Foucault, 'biopolitics' refers to regimes of power premised on the management of life, or populations, whose productions can be deployed as an organic arsenal.

Performance Research 13(3), pp.18-30 © Taylor & Francis Ltd 2008
DOI: 10.1080/13528160902819281

towards the common misfortunes and injustices of life, easily witnessed in the streets of New York City, to Jesus' early stumblings. The priest comments, 'unemployment makes people feel worthless and unproductive, and takes a terrible toll on the stability of the individual and the community.' In turn, the public response, in a chorus of unrehearsed voices, confirm the meditation:

Compassionate friend,
We lose interest in your journey to justice
As we worry about our own needs.
Similarly, we dismiss the unemployed
As unfortunate but having too large a problem for us to address.
Give us courage to respond to the challenge of unemployment.

In a ritual rehearsal of Mary's mercy that, in its liturgical genealogy, is a kind of prelude to the classic image of Michelangelo's Pietá (resplendently housed in the Vatican), the fourth station was enacted outside the low-lying cinder-block church of San Emérico, incongruously juxtaposed against the metal sprawl of the Keystone electrical industrial complex that borders the East River. Cordoned off by walls and walls of barbed wire, in this social context the passion play confirms the environmental health hazards that plague low-income communities.

Noble hijo de María,
El testimonio paciente de tu madre nos ha debido enseñar
Desde hace mucho tiempo a ser tolorantes.
Sin embargo, todavía juzgamos a grupos enteros
En base a estereotipos o a incidents aislados.
Envia a tu Espirítu para que desvanesca el prejuico
Que nos impiede ver las virtudes de los demás.

Whereas lower Manhattan's historical Trinity Church enjoys the pastoral refuge of a garden and a tree-dotted historical graveyard (an Anglican stronghold whose parishioners at Broadway and Wall Street have included George Washington in 1789 and post-11 September, 2001, rescue workers), San Emérico is entirely surrounded by the cheap expanse of an asphalt parking lot. This is the socially constructed space - of economic, environmental and cultural marginality - that frames the remaining ten stations of the Loisaida passion. Quite obviously, to say that this performance was 'site specific' points to the uneasy inadequacies in traditional conceptualizations of an avant-garde that has always been rigorously diverse in form and content, and in cultural location (Mercer 2005; Harding and Rouse 2006).

At the fifth station, where Simon helps Jesus to carry the cross, there is a change in the cast: we see another 'Jesus' step in to relieve the first en/actor of the son of God, sharing a burden of ritual representation in tandem with the theme of Simeon Cirineo's friendship that also achieves the theatrically 'experimental' collectivity of solidarity: between en/actors, between (in this case) men, and between publics. Each stands, in the Viacruces, for that Christian traffic between the embodiments of the quotidian (profane) and the sacred. The collective response confirms, 'Diligent son of a carpenter, we long for Simon's opportunity to assist you with the cross.' Three blocks south, at the seventh station (where Jesus falls for a second time), parishioners call for 'a true understanding of poverty, so that we may one day overcome it'. Each scripted response is prompted by a parish priest's spoken commentary, whose oration here insisted, and with particular force:

We have all gone astray, He hears the guilt in us all. The poor have the same problem, which is not merely a lack of resources, but a denial of the full participation in economic and political life, the ability to be in life, which is fundamental to human dignity.

This is the kind of language that activists and scholars currently recognize in the strengths of Amartya Sen's 'capability approach' to human rights practices and theory, a conceptualization of the human subject in the context of collective social well-being that addresses human capabilities not through the model of possessive individualism and fair competition (usually driving human rights discourse as a paradigmatic instrument of liberalism) but through collectivities and communities - a model that begins to grasp the radical collective goals of the Loisaida performance.[3] To think in terms of the capabilities approach privileges a conceptualization of life as requiring (but as Sen insists, is not limited to) education, healthcare, bodily integrity, choice and political participation. In the context of the mainstream US political tradition, this model can be useful in order to raise 'awareness that you do not secure the necessary ingredients of democracy without at the same time focusing on material issues such as health care and the provision of universal primary and secondary education (Nussbaum 2008: 22). At the seventh station, in sum, the technique of passion play performance, the content of the commentary and the public response emphasize the respective values of collectivism and solidarity.

At the performance of the eighth station, where Jesus Christ speaks to the weeping women in Jerusalem, the priest laments that the people of Loisaida 'no tienen ayudo, ni medico, ni tratamiento prenalidad / they have no help, no medical assistance, or even prenatal care.' At the same time, Loisaida women swept to 'Jesus' in tears, begging mercy for their own families: 'nosotros tambien nos preocupamos, por los ninos que ya nacen en miseria, y que viven en medio de terribles privaciones / we are also worried about our children, born in misery, and who live with terrible burdens.' The Loisaida mapping of the passion was punctuated once again by significant site-specific visitations, identifying the observation of location itself as an act of witness, and beseechingly so at the ninth station (where Jesus falls for the third time): denouncing drugs and violence on a 9th Street block known for its crack and heroin dealing. The priests told everyone, there in the street, that 'the state is very concerned with this block. This is the beginning of a major campaign that we are launching with the police. We want everyone to take notice, y que no vamos a seguir tolerando / and we are not going to tolerate it anymore.' More, the priests warned that they knew who was selling narcotics, '¡quienes venden drogas, tenemos nombres / we have names!' in an effort to reclaim the public health of the Loisaida community. The mural at 9th Street and Avenue B reads, 'La Lucha Continua', and graffiti screams *'DIE YUPPIE'*.

The Loisaida passion turned west up 9th Street to stage the crucial thirteenth station (the Crucifixion) on a grassy mound in Thompkins Square Park, with the solemn quiet due to the performance of the emblematic sufferings of Jesus Christ, nailed on the cross in a manner of execution that was once so common it accrued no particular symbolic importance other than the banality of punishment by death, until Christianity claimed it as the symbol of resurrection for petty thieves and the 'son of God' alike, unified in the final reckonings of compassion. Again, the cast of 'Jesus' en/actors changes. This third embodiment (re-enactment) is stripped of most of his clothes, revealing a body - a mortal person - with significant scars, signs of a deep cutting that tore from the heart all the way back through to the spine. These are not special effects but the grisly marks of global/local violence, and survival. The lean and scarred body speaks the lines everyone expects: 'forgive them, for they know not what they do.' In view of this suffering body, the priests condemn 'greed, prejudice and disinterest' and the destructive

3 Amartya Sen remarks, 'The problem is not with listing important capabilities but with insisting on one predetermined canonical list of capabilities, chosen by theorists without any general social discussion or public reasoning. To have such a fixed list, emanating entirely from pure theory, is to deny the possibility of fruitful public participation on what should be included and why ... public discussion and reasoning can lead to a better understanding of the role, reach and significance of particular capabilities' (2004: 77, 81). Martha Nussbaum's list of ten 'central human capabilities' include Life, Bodily Health, Bodily Integrity, the Development and Expression of Senses, Imagination and Thought, Emotional Health, Practical Reason, Affiliation (both personal and political), Relationships with Other Species and the World of Nature, Play and Control over One's Environment (both material and social) (2008: 23).

'approach to farmworkers, the poorest paid migratory field workers who are particularly susceptible and who need housing and health and pay'. In turn, the public response, its penultimate chorus, delivers this sound:

Incarnate wisdom,
so often we feel confused by the talk of political leaders.
We are left uncertain about how our nations should respond
To the crisis of world poverty.
Share your wisdom with us so that we might act more effectively
On behalf of the world's neediest people.

At the close of just over two hours of the passion play, the troupe finally enacted the fourteenth station, offering prayers of devotion to the fallen Jesus who would rise again (Easter), inside the broad halls of Santa Brígida. The priest talks about 'the life-giving death of Jesus Christ our saviour' and the need to 'reduce poverty in the third world'; he asks the congregants inside the church to think about 'what we can do together, how can we form cooperation, directed to real social needs?' and finally suggests particular attention to the community of Mexican migrants up at 14th Street and First Avenue, who are hungry.

La Catorce Estacion: Jesus es colocado en el sepulcro.
The Fourteenth Station: Jesus is laid in the tomb.

Respuesta: Libertador poderoso,
Nos llamas de la muerte a la vida, de la independencia a confiar en ti.
Libranos del miedo que evita la reconcilacion de los enemigos.
Permite que cambie nuestra vision de la economia
Para construir tractores y viviendas
En lugar de tanques y misilis.
Amén.

Response:Almighty liberator
Call us from death to life, to the independence to confide in you.
Free us from the fear that evades reconciliation between enemies.
Help us to change our economic vision
To create tractors and homes
Instead of tanks and missiles.
Amen.

With this radical re-working of the passion, Loisaida clearly understood itself as a critical participant in the world, contesting military expenditures, agricultural policy, housing issues and human rights principles alike. The performance converted New York City (and the nation) to the liberation-theology interpretations of the passion by mapping another kind of liturgy: enacting a communal response to the everyday sacred of its people and their public histories. Speaking into hand-held microphones, priests insistently matched the time of Jesus to the time of 1998: where Jesus assumed the cross, fell for the first time and was crucified, so too do the people of Loisaida bear its cross of poverty, fall to unemployment and endure the crucifixion of indifference to suffering. The writing of an unmistakably anguished Jesus doubles, in this passion, for Loisaida impoverishments, as the performance details community afflictions such as joblessness, homelessness and drug addiction. As a dramaturgical script, this passion performance theatricalizes what has become that familiar ritual - of local opposition to distant machineries of power - and in so doing expresses a popular rejection of modernity's rampant failures. This passion, moreover, asserts the authority of secular life (the profane) in the labour towards 'spirit' - exhibiting in 1998 (and in the U.S., no less) the seemingly spent tradition of Latin American liberation theology, which since the early 1960s has presented the theological legitimacy of the question of social liberation through such biblical instances as Isaiah 61: 'Our mission, as that of Christ, consists in giving the good news to the poor, proclaiming liberation to the oppressed.'

Liberation theology, sometimes called a 'socialist catholicism', is concerned with social justice, human rights and poverty. In Philip Berryman's summary it is 'an interpretation of

Christian faith through the poor's suffering, their struggle and hope, and a critique of society and the Catholic faith and Christianity through the eyes of the poor.' Whereas the Vatican II (Second Ecumenical Council of the Vatican, 1962-5) re-affirmed the importance of church hierarchy as part of its project of church renewal in the modern world, the interventions of such a 'popular church' is characterized by collectivist and bottom-up social organization of church practices. As Enrique Dussel explains, 'We must realize that we are involved in a passage through history towards liberation' (1976: 143), which is an *existential praxis*, or process in time, working towards the freedom of life:

> *A new historical human being beyond that of the relation of domination that oppresses all the underdeveloped peoples, beyond all historical mankind, is the final Kingdom of God. The struggle for liberation, the leaving behind the land of colonial*

servitude, is the hope of salvation. It signifies a new era. As a sign of God's grace it falls on us to be living at this time, and we are part of the adventure of seeing the dawn. (1981: 253)

In the passion performance, Catholicism is not only performative in its traditional liturgical routines, it radicalizes an idea of the creative spirit of life itself, dedicated to participating in the historical processes.[4] Still, at the 1979 meeting in Puebla, Mexico, Pope John Paul II maintained that 'this conception of Christ as a political figure, a revolutionary, as the subversive of Nazareth, does not tally with the Church's catechisms.'

In its social mappings of public space, the 1998 passion performance abandons the more common convention of dramatizing established scripture for liturgical observances and staged performances, which historians suggest have in some way been present since the sixth century, under Pope Leo the Great (Young 1910). Documents of Latin passion play scripts date to the early twelfth century (Sticca 1967), whereas recovered textual evidence of passion plays using vernacular language date at least to the fourteenth century (Frank 1920). As part of a week-long culmination of religious observation marking the forty days of Lent, and following the observation of Holy Thursday which commemorates the 'last supper' of Jesus with the Apostles, 'the script' of the 1998 Loisaida passion replaces the traditional procession of Holy Friday (sometimes Good or Great Friday) that prepares for the celebrations of Easter Sunday. In the context of the U.S., the Roman Catholic Loisaida passion clearly departs from the for-profit model characterizing the elaborate and spectacular Protestant evangelical stagings since the late nineteenth century (Alexander 1959; Chansky 2006). The Protestant Reformation (begun in 1517), of course, was a process that opposed what might have been considered the false idolatries of such ritual exhibition (in 1549, for example, the Synod of Strasburg opposed all religious plays), but by the nineteenth century, a revival of interest in the passion play has yielded what has become the largest outdoor dramatic (i.e., scripted) performances of the twentieth-century United States, with the passion plays of Eureka Springs, Arkansas, and Glen Rose, Texas, advertising the cause of staged ritual and religious pilgrimage as sites for regional tourist attractions.

The 1998 Loisaida passion performance is distinctive in several ways. The utopias of community and the labours of sacrifice conjured in the passion illuminate reckonings of exchange rituals that are at the heart of Christianity. If morality plays present dramatized allegorical lessons about how to purchase redemption (*Everyman*, circa 1485), and mystery plays rehearse liturgy as the ritual representation of Christian belonging, the 1998 Loisaida passion performance reinforces a Christian interest in accounting for embodiment and the constitution of social space, and in explicit relationship to the cost of restitution in a counter-public spectacle (Negt and Kluge 1993). 'Interest', of course, signals, spiritual, ideological, as well as financial reckonings. In his study of Everyman, John Parker argues that the medieval play demonstrates the proverbial 'everyman' struggle to keep his life, offering money and earthly goods to stave off the inevitabilities of mortality, when only redemption through Jesus Christ (and by extension, the lettings of Christ's blood) will do. Christianity thus traffics in a 'radically alien form of exchange' that positions 'humans as themselves a form a currency' (99). If Jesus Christ pays for all mortal debt with his blood, Parker reasons, Christ's blood can be seen as God's chosen form of currency; casting God's will in the traditional role of ransomer, and redeemer, while Jesus Christ's offerings provide the mediations of mercy: 'Everyman becomes a bequest to Christ to satisfy his spiritual debt only when his blood and flesh have been ... made into payment' (101). This apparent economistic analysis clarifies the exchange relationship between mortality and the sublime, which, while not ecumenically preferred, was not exactly

4 It was in the 1960s, of course, when Brazilian Augusto Boal first developed the Theatre of the Oppressed at the Arena Theatre. In 1967, Glauber Rocha's manifesto for the radicalization of the Brazilian Cinema Novo ('The Tricontinental Filmmaker: That Is Called The Dawn') used language similar to Dussel's. (Randal and Stam 1982: 76).

uncommon (if yet untoward) in its own time. Indeed, the performance of Everyman's accountings exemplify what William Eggington calls 'the ephemeral space of medieval theatre' (1996: 10), wherein ritual enactments negotiate the stuff of human accountings, at once clerical and 'civic' concerns. This is the management of property exchange relationships, if you will, between life and death. Rather than repress it, Liberation Theology seizes the exchange relationship between the sacred and the profane as the opportunity of Christian performance. Treading the ground between the 'spectacular exhibition of power [and] the bureaucratic representation of it' (19), such performances acknowledge the relationship between signified and signifier in mimetic practices that promise to unravel the authorities of each. To put into play allegorical characterizations of Good Deeds and Death, as God's ransomer (amidst the undercurrents of simony) - like the allegorical exhibition of Jesus Christ and 'believers' (amidst the military-industrial complex) - is to reveal spiritual and civic ideologies as administrative rationales that traffic in corporeality. The 1998 Loisaida passion performance, in sum, maps reminders about the price of the ticket. (Baldwin 1985)

Ngugi wa Thiongo specifies performance space as a combination of fields including the totality of its external relations, and the relationship to time, elaborating a conflict between performance and the state as a contest in the enactments of power. 'The main arena of struggle is the performance space: its definition, delimitation and regulation' (1997:12). In the US context, any public event, whether a passion play procession or a political demonstration, requires the permission - and the armed presence - of the police state, effectively criminalizing the public in the interest of 'security'. In March 1998, the contest for the enactment of power is clear: while parish priests thank the NYPD for their traffic control assistance, they also harangued them for the brutalizations and impoverishments of its people, rallying the community through the passion play to march for its own cause. They did so through a performance of Christian time that scripted spatial genealogies of community politics, as well as its regulations and punishments by the state. To think the continuities and change in time is, indeed, to think history. Eric Alliez defines 'capital time' as 'the nexus between temporalization, capitalization and subjectivity', in which 'the most aberrant movement has become the everyday itself, the daily mastery of time' (1993: xvi). The Loisaida passion play performance reproduces an observation of 'capital times', not because there was any commercial media coverage of the event (there wasn't) but rather because the biopolitics of the history of capital itself were the subject of this rather more *catholic* (universalizing) performance.

As per Heidegger time, like space, 'is that which is let into its bounds'(1971). What histories belie this nexus of capital time? After Operation Bootstrap (1948) annexed Puerto Rico to the centralizing regimes of US global production strategies, and with the deindustrialization processes of US cities in the 1960s, the spatial and technological transformations of economic activities sank Lower East Side residents into deeper poverty[5]. As citizens of the US since 1917, and of the Commonwealth since 1967, Puerto Ricans had long resided in New York City, even though, in 2008, they are still referred to as 'foreigners'.[6] In the 1970s stagflation, property values plummeted so low that owners abandoned buildings to disrepair. The popular phrase 'burn, baby, burn' is rooted in the practice of absentee landlords in this period, who burned their properties in order to reclaim some kind of value out of their assets without regard for the low-income tenants who lived in, and were killed in, these buildings. The Young Lords Party spearheaded several local Nuyorican institutions, including El Bohio Cultural and Loisaida Folklórico, and in what has become known as the case of 'Garbage activism', the Young Lords organized the pickup and removal of street-side trash, a public service the nearly

[5] A topic notoriously narrated by anthropologist Oscar Lewis, in *La Vida: A Puerto Rican Family in the Culture of Poverty - San Juan and New York.* New York: Random House, 1965.

[6] On 6 February 2008, the New York Congressional Delegation press release states: 'Congressman José F. Serrano (NY-16) and Congresswoman Nydia Velásquez called on Representative Ginny Brown-Waite (R-FL) to apologize to the Puerto Rican community after the Florida Republican put out a statement saying that as "foreigners" Puerto Ricans did not qualify for tax rebates offered through the economic stimulus package' that President George W. Bush petitioned for a declining economy.

bankrupt city only selectively provided to its taxpayers. The city, for its part, demolished some buildings as health hazards and left others standing, leaving a pock-marked pattern of blocks. In the wake of disappeared buildings emerged a culture of Nuyorican *casitas* and impromptu gardens: part nostalgic recreations of a better Puerto Rican landscape, part mini-agricultural opportunity and part local dispossession. Today some of these lots grow enormous trees and others are junkyards; some casitas yet remain. Begun in 1983, Operation GreenThumb gardens (also Giuliani-era neighbourhood development projects heralding gentrification), positively flower against the Bloomberg-era rebounding real estate development designs on previously 'dangerous' addresses.[7]

The 1998 passion play performance works out Hamlet's familiar observation (and Dipesh Chakrabarty's argument) that 'time is out of joint', or that modernity's 'historical time is not integral ... it is out of joint with itself'.[8] In the 1990s, Loisaida churches continued efforts at activism despite the global financial freeze on progressive politics. While 'Loisiada' residents fought city-sponsored, Giuliani-driven real-estate development projects that 'resurrected' abandoned buildings and dislocated lower-income and Nuyorican communities, the March/April 1998 *The Catholic Worker* (an organ of the Catholic Worker movement founded by Dorothy Day and Peter Maurin) ran front-page stories on 'Downward Mobility' and excerpts from Martin Luther King, Jr's last Sunday morning sermon, delivered 31 March 1968 at the National Cathedral (Episcopalian) at Washington D.C., which was in fact Easter (Passion) Sunday, four days before he was assassinated:

> It is no longer a choice, my friends, between violence and nonviolence. It is either nonviolence or nonexistence, and the alternative to disarmament, the alternative to a greater suspension of nuclear tests, the alternative to strengthening the United Nations and thereby disarming the whole world may well be a civilization plunged into the abyss of annihilation, and our earthly habitat would be transformed into an inferno that even the mind of Dante could not imagine.

Similarly, the passion play performance articulates a community that is as bound to the authority of Papal law in Rome as it is (again) to the administrative failures of public policy. This performance was saturated with protests of the everyday 'real', where the staging of Christian time initiates a stalling of modernity's global time because it injects reconnaissance to local time by way of the ritual religious play that reinstates both, simultaneously. No wonder, medieval theatre historians notice, the formalities and elitist apparati of the now-conventional 'theatre', properly understood, were developed and supported by the courts - to usher in aesthetic regimes more obedient to the crown and church alike. The performance recalls, with sudden clarity, motivations for taking seriously Paul Virilio's claim that in artistic work today 'we return to the profane body'.

Ghosts, Valladolid (1550)

Notwithstanding the 1549 protestant prohibitions against religious plays in Germany, in Catalonia (northeastern Spain) the first documented passion play dates to 1538, when Christian rituals of exchange were not any more 'abstract' in the sixteenth century than they are in the twenty-first. The 'unification' of Christian Spain, heralded by the marriage of Ferdinand and Isabella, fuelled the *re-conquista* of Spain and the *conquista* of the New World. As enemies of Christ, Africans were, historian Darién Davis points out, denied the exchange conventions of Christianity and made into objects of the lesser remittance demands of money (slaves), as opposed to the faithful, who paid through the recognition of the blood of Jesus Christ as holy currency.[9] In *Medieval Slavery and Liberation*, Pierre Dockés defines slavery as a state of reprieve, to be 'in a state of suspense, as it were', or to be 'held to be dead', as 'the fate of the debtor whose creditor forgoes the right to exact his due in blood'(1982: 5). This is resource management

7 See www.notbored.org/gardens.htm. Accessed 27 February 2008.

8 Re-reading subaltern studies, Dipesh Chakravorty explains that the provincializing project 'refers to a history that does not yet exist ... the idea is to write into the history of modernity the ambivalences, contradictions, the use of force, and the tragedies and ironies that attend it ... As I have said before, it is impossible within the knowledge protocols of academic history (2000: 42-5). In the case of the Loisaida passion play performance and in the liturgies of Christianity, too, time is not integral: 'Christ yesterday and today / The beginning and the end/Alpha and Omega / All time belongs to him/and to the ages: / To Him the glory and the power / Through every age forever / Amen'.

9 Darién J. Davis writes, 'in 1455, Pope Nicolas V gave the Portuguese the right to reduce to slavery the inhabitants of the southern coast of Africa who resisted the introduction of Christianity, thus in theory becoming the enemies of Christ. As a consequence, the Iberians began a modest slave trade on the Western coast of Africa .... The Portuguese set up factories, or trading posts, to deal with local middlemen and tribal chiefs. Africans contributed to the diversity of Iberian cities such as Seville and Lisbon, which were already inhabited by Jews, Arabs and Christians. Small communities of Afro-Iberians thus emerged' (1995: xi).

of profane goods - not altogether foreign to the human-rights discourse of Sen's deliberations about human capabilities and collectivities, or to the Loisaida passion play performance. The difficulty of distinguishing the slippages between the two logics was the topic of a February 2005 Public Theatre, New York, performance about the discussions at Valladolid in 1550, when the Franciscan Friar Bartolomé de las Casas was recalled from Nueva España to the city of Valladolid to submit his testimony against that of Ginés de Sepulveda for a closed review that was called by the Papal Legate. The Public Theatre featured the play *The Controversy of Valladolid* as part of a small series of theatre works curated to investigate the realpolitics of religious governance in contemporary global cultures, at once defined by commercial and military interests; not an unsalubrious project.[10] Letters written by Bartolomé de las Casas and Gines Sepúlveda supply the only trace of the specific language that was exchanged at the tribunal, allowing us to guess at how the hearing unfolded, and what analysis of *thought* occurred there. In the staged rendition, indigenous 'Others' are called into the room for the benefit of the Legate's observation, not as witnesses with the capacity to testify but as bestialized objects of evidence (species samples) for his Holy interrogation.[11] Implicit in the background is the slaughter of Spanish conquest in the New World. De las Casas argues, Sepúlveda argues, and the Papal Legate decides: it is a tribunal, where the violence of words authorizes all violences.

The curious thing about the 1550 tribunal is that the status of indigenous peoples had been decided before, and many times over, although the play never mentions it. Theoretically, the conventional exchange relationship between mortal subjects and God (including the costs of salvation) had been extended to the New World. Locally, the *Siete Partidas* (Book of Laws) of Alfonso X, King of Castile y Leon (1252-84), is a civil code that addresses the principles for the proper relationships between people and specifies its rejection of slavery.[12] The 1462 Papal declaration by Pope Pius II condemned slavery. A 1537 Papal Bull *Sublimus Deus* (by Paul III) declared 'Indians' rational beings with the capacity to understand and receive Christian faith, whose lives and property should, therefore, be protected (de las Casas 1992: 6). On this basis the 1542 *Reyes Nuevas* again explicitly outlawed Indian slavery. For his attempts to enforce the law, Bishop Antonio de Valdivieso of Central America was assassinated in 1550 on orders from the Governor of Nicaragua. This is just a smattering of evidence; there is plenty more for the archivist to examine. Paraphrasing Sepúlveda's *De la justa causa de la Guerra contra los indios*, the speech in the script elaborates the old argument: 'Aristotle said it best: some human species are made to rule over others. I do not deny their human condition. That would be absurd. I simply say that they are at a lower stage of this condition.' The dubious distinction intends to justify enslavement, a rationalized management of profane life that, within the strict economy of Christian logics, at least, suspends human goods from the exchange economies of Jesus Christ's blood, effecting total alienation and dispossession:

> Reprieved from death, set apart, isolated from every community, hence dead, alien to citizenship, homeless, a stranger to himself, the slave is no longer a man, but a member of a subhuman species located somewhere between women and animals: we have only to recall Aristotle's well-known classification, in which the slave was an animate instrument, like an animal, but more efficient than an animal, being endowed with speech. (Dockés 7)

In part, Sepúlveda's task was to justify Spain's implied insubordination to Rome; Sepúlveda argued about rationalized structures of resource management, in order to imagine war and slavery as theologically coincident with Papal authority.[13] It is important to note that Sepúlveda's arguments for conquest built upon Francisco de Vitoria's *Just War Theory*, elaborated from Cicero and Aristotle before, to confirm the exploitation of the Americas for

[10] The English version that appeared at the Public Theatre was translated by Richard Nelson and directed by David Jones. The play was written for television in 1992 by Jean-Claude Carriére and subsequently adapted for the stage in 2004. Its limited use of the theatrical space did little to extend the gaze of the audience beyond the confined close-up of the televized dialogue, a technique that narrows interpretational regimes to the atomized perspective of the individual he-said/ he-said courtroom melodrama.

[11] Much has been ventured on the representation of the ethnographic subject and the exhibition and performance of objectified persons (see Fusco 1994). As a result of the violences in the absolute conflict of colonial war, Enrique Dussel clarifies: 'I cannot condone dominant elites in Latin America or Spain who continue speaking of the meeting between two worlds' (1995: 55).

[12] In 1789 Alfonso was included among the twenty-three lawmakers depicted in the US House of Representatives (at the same time that slave labour was used to build the US Capitol).

imperial interests. To this day De Vitoria is honoured as a founder of international law; Sepúlveda figures as his minor apprentice; recalling Nietzsche's always astute apprehensions about 'priestly asceticism' as 'the best instrument for yielding power ... the "supreme" license for power' (1967: 47).

This stands in contrast to de las Casas's concern with the case of indigenous subjectivity and to his aim to change the abuses of the *encomienda* system. What facilitated the plea for the rights of the human subject was his deployment of the Christian exchange economy. History suggests there are several faces to the currency of Jesus Christ. Thereafter, the appointment of de las Casas as Bishop of Chiapas in 1554 did little to stem King Philip II's drive (1556) for control through the twin powers of military conquest and the increasingly stringent laws of Censorship (1558) that regulated criticism as heresy. Two years after Valladolid, of course, the publication of *The Devastation of the Indies* by de las Casas further documented Spanish violences against indigenous peoples, which had not ceased. Hoping to demonstrate the ills of this abjection, the play reviews what can become the arbitrary biopolitics of religious rule. What rubs off as the dusty anachronisms of the Valladolid controversy instead returns as a politics of recognition: where spectres of exchange between the profane and the sacred continue to animate political 'spirit'.

Ghosts, 2005 (Take 1)

In the autumn of 2004, the church of Santa Brígida on Thompkins Square Park was closed as a result of financial strains in the Roman Catholic Church; María Auxiliadora would soon follow. The 25 March 2005 Good Friday performance proceeded with two of the remaining neighbourhood churches but showed no signs of the local activism or liberation theology that so defined the earlier passion play performances as political as well as avant-garde projects. For a direct textual comparison, the response text for the fourteenth station performed in the same neighbourhood on 25 March 2005 read:

Señor Jesús,
Guíanos a través de las etapas de nuestras vidas,
Desde nuestro nacimiento a nuestra madurez
Y hasta nuestra última jornada hacia Dios.
Esto to lo pedimos, Señor y Salvador nuestro,
Que vives y reinas por los siglos de los siglos,
Amén.

Dear Lord,
Guide us through the stages of our lives
From our birth to our death, until the day
Of judgment when we come before God.
This is what we ask of you, Lord,
Who lives and reigns in the time of the ages,
Amen.

The disciplining of the text back to its more conventional content is striking, returning the event to the status of a more strictly institutionalizing religious ritual. In 2005 the Passion reverts to a traditionally paternalistic Christianity. The limits of the traditional liturgical theatres of the Passion throw into the relief Baz Kershaw's observation that 'performative excess untrammelled by theatre is much freer to create new domains for radically democratic practice (1991: 85), which the re-writings in the 1998 passion play perform, reminding audiences of the difference liberation theology performance once made in this neighbourhood. That is to say, the continuity and change in the Loisaida passion performances announce that we are always again 'medieval' in the contemporary moment, and not in any simplistic rehearsal of violence and the sacred.[14]

(Take 2)

24 March 2005 was also the twenty-fifth anniversary of the assassination of Archbishop Oscar Romero in El Salvador. The country prepared for massive commemorations. Despite the refusals of Rome, Romero has been evangelized by the people of El Salvador, who honour him for finally remembering that controversial idea: that the poor matter for a

[13] Then as now, rationality is said to be that species-differentiating property of the human which, in the sixteenth century at least, made it possible for the human species 1) to have a relationship to other people and their social organizations and 2) to have a relationship to the Christian God, heretical and otherwise. In Coetzee's ventroloquism, 'The fact that animals, lacking reason, cannot understand the universe but simply have to follow its rules blindly, proves that, unlike man, they are part of it but not part of its being: that man is godlike, animals thinglike'(2003: 67).

[14] René Girard, a conservative Catholic in the league of US Supreme Court Justice Antonin Scalia, links violence with the sacred as a necessarily sacrificial adjudication of Christianity. For Girard's position on Pope Benedict XVI (Cardinal Ratzinger's) theology, see 'Ratzinger is Right' (2005).

pedagogy of the oppressed. The famously anti-communist Polish Pope, consecrated in 1976, did not like Liberation Theology or its links to Marxist thinking in Latin America. This position translated into clear Vatican policy in the Americas: alliance with the military and the power elite. At the 1979 conference in Puebla, Mexico, Pope John Paul II declared that the Catholic Church should not be involved in politics. The Pope went so far as to condemn Romero, among others, for their parts in witnessing the popular protests and the massacres of their communities. It is speculated that this amounted to Romero's death warrant: if the military needed to act against those who resisted them, including the clergy, the Vatican would not interfere. In 1979, liberation-theology bishops nevertheless sent Romero a letter of solidarity:

> We know that the Lord placed upon your shoulders the pastoral responsibility of the Archdiocese of San Salvador at the time in which the chastisement, the veritable persecution, began. In the midst of all this, accused and defamed along with those who search for ways of justice, you have remained steadfast, knowing that you have to obey God rather than men.

A few months later, Romero was shot down by a military squad while giving Mass on Holy Thursday 1980, the day before the Good Friday passion play processions. He did not make it to his Easter sermon. In El Salvador, the national genocide was administered through a US-backed military regime, part of a Kissinger plan to manage Central and South American markets for US interests. Opposing this regime was apparently against Vatican policy. In that case, Sister Pamela Hussey declared in a 2005 interview with the BBC about the anniversary of Romero's death, 'I am a Communist and a Marxist.'

(Take 3)

Throughout Holy Week, beginning 24 March 2005, Pope John Paul II has been slowly and publicly dying or, in the words of the faithful, 'serenely abandoning himself to God's will'. Journalists discuss the contradictions of the secular age when the Pope, born Karol Wojtyla in 1920, leaves the hospital, refusing the surrogate life of medical technologies in favour of 'natural death'. This act of dying confirms and performs the exchange economy Parker points out, repeating the productivity of biopolitical life for the Christian faithful, and inscribing the prosthetic/Real death of God, all in the hungry eye of the global media. On Saturday 2 April, the Pope dies. The wish of some in Poland that his heart be buried in Poland is denied (in the past, Popes were often embalmed). In his tenure, the Pope negotiated profane/sacred exchange economies by supporting campaigns 'for life' (against reproductive choice and safe sex), opposing US President George W. Bush's campaign for war in Iraq, and affirming that the principle of private property 'must lead to a more just and equitable distribution of goods ... and if the common good demands it, there is no need to hesitate at expropriation itself, done in the right way'. The Archbishop of Canterbury attended the Pope's funeral, for the first time since the Church of England broke from the Papacy in 1534.[15] Cardinal Ratzinger, who as head of the Congregation for the Divine Doctrine of Christian Faith previously condemned liberation theology, now became Pope Benedict XVI.

[15] US and other heads of state attended the funeral services, including Russian Prime Minister Mikhail Fradkov, Iranian President Mohammad Khatami and Israeli President Moshe Katsav.

Ghosts, 2008

The theatricality of the 1998 passion play ritual showed imaginative and inventive prowess as activist and avant-garde performance. Ten years later, only the parish at San Emérico remains. Marvin Carlson has suggested that theatrical performance is 'an activity deeply involved with memory and haunted by repetition' (2001: 11), to the extent that

> the ghost has a greater performative visibility than the body it haunted ... transcending the body of flesh and blood, this other body consisted of actions, gestures, intentions, vocal colors, mannerisms, expressions, customs, protocols, inherited routines, authenticated traditions – 'bits'. (11)

The 1998 passion play performance was not a simple activity of surrogation but a haunted re-invention of the logics inhabiting liberation-theology approaches. It is with the permission of Roland Barthes's course on 'The Neutral', that thing of ambiguity that outplays or baffles paradigms, that I have taken up performances of Roman Catholic paradigms of secular or profane life, ideas that have become apparent - or better, baffling - in performances that inconsistently utter the words 'secular', 'political' or even 'profane' yet invoke their uncertainties to achieve theological effects.[16] No doubt, to view institutionalized religion as a political organization, and its doctrine as law subject to critique, is an unapologetically secular assertion, in league with Edward Said's closing contemplations that 'the humanities concern secular history, the products of human labor, the human capacity for articulate expression' (2004:15). At the same time, secular universalism struggles to account for difference in rationalized narrative orders that, as Chakravorty suggests, remain inadequate to the radical diversities of life itself. In a country mythologically premised on the 'right' to freedom from religious persecution, issues of religious performance command the disciplining effect of *neutrality* in all public discourse on the topic in the United States, often eliciting the silence of the sceptic. (Everyone, meanwhile, has permission to disparage lawyers.) For the secular humanist, the theatre and performance scholar, the practice of research is investigatory: not simply to neutralize norms and normativities, as they are codified into laws that can be understood as regulatory if not simply repressive, but more importantly to explore performance as a preeminently productive arena of aesthetic and social interpretation. With this writing I have addressed the question of how figurations of these definitively Roman Catholic exchange rituals trace what haunts them for thinking secular life, because, as Gilles Deleuze wrote in his final essays, '*a* life is everywhere' (2001: 29) and many discourses - structures of feeling, theocratic and otherwise - are poised to claim the living.

REFERENCES

Afary, Janet and Anderson, Kevin B. (2005) *The Seductions of Islamism: Gender and the Iranian Revolution*, Chicago: University of Chicago Press

Alexander, Doris M. (1959) 'The Passion Play in America', *The American Quarterly* 11(3): 350-371.

Alliez, Eric (1993) *Capital Times*, Minneapolis: University of Minnesota Press.

[16] Barthes writes: 'I define the Neutral as that which outplays *[déjoue]* the paradigm, or rather I call Neutral everything that baffles the paradigm. For I am not trying to define a word; I am trying to name a thing: I gather under a name, which here is the Neutral' (2005: 6). He continues, 'We might perhaps say that the Neutral finds its feature, its gesture, its inflection embedded in what is inimitable about it: the smile, the Leonardian smile analyzed by Freud ... in which the mark of exclusion, of separation, cancels itself. To the gesture of the paradigm, of the conflict, of the arrogant meaning, represented by the constraining laugh, the gesture of the neutral would reply: smile' (195). Thus for Barthes the practice of the neutral is *to smile*, 'releas[ing] the prisoners: to scatter the signified, the catechisms' (xiii).

Artaud, Antonin (1958) *The Theatre and Its Double*, New York: Grove Press.

Baldwin, James (1985) *The Price of the Ticket: Collected Fiction and Nonfiction, 1948-1985*, New York: St. Martin's Press.

Barthes, Roland (2005) *The Neutral: Lecture Course at the Collége de France (1977-78)*, trans. Rosalind E. Krauss and Denis Hollier, New York: Columbia University Press.

Carlson, Marvin (2001) *The Haunted Stage: The Theatre as Memory Machine*, Ann Arbor: University of Michigan Press.

Casas, Bartolomé de las (1992) *The Devastation of the Indies*, A Brief Account, trans. Herma Briffault, Baltimore: Johns Hopkins University Press.

Chakrabarty, Dipesh (2000) *Provincializing Europe: Postcolonial Thought and Historical Difference*, Princeton: Princeton University Press.

Chansky, Dorothy (2006) 'North American Passion Plays: "The Greatest Story Ever Told"', TDR 50(4): 120-145.

Coetzee, J. M. (2003) *Elizabeth Costello: Eight Lessons*, London: Secker and Warburg.

Davis, Darien J. (1995) 'The African Experience in Latin America: Resistance and Accomodation', in Darién J. Davis (ed.) *Slavery and Beyond: The African Impact on Latin America and the Caribbean*, Wilmington, Delaware: Scholarly Resources, Jaguar Books on Latin America.

Deleuze, Gilles (2001) *Pure Immanence: Essays on A Life*, New York: Zone Books.

Dockés, Pierre (1982) *Medieval Slavery and Liberation*, Chicago: University of Chicago Press.

Dussel, Enrique (1976) *History and Theology of Liberation*, New York: Orbis Books.

Dussel, Enrique (1981) *A History of the Church in Latin America*, Grand Rapids, MI: Eerdmans.

Eggington, William (1996) 'An Epistemology of the Stage: Theatricality and Subjectivity in Early Modern Spain', *New Literary History* 27(3): 391-414.

Foucault, Michel (1977) *Discipline and Punish: The Birth of the Prison*, New York: Vintage Press.

Frank, Grace (1920) 'Palatine Passion and the Development of the Passion Play', *Modern Language Notes* 36(5):193-204.

Fusco, Coco (1994) 'The Other History of Intercultural Performance', TDR 38(1).

Girard, René (2005) 'Ratzinger is Right', PostNational Literature (Summer), <http://www.digitalnpq.org/index.html>.

Harding, James and Rouse, John (2006) *Not the Other Avant-Garde: The Transnational Foundations of Avant-Garde Performance*, Ann Arbor: University of Michigan Press.

Heidegger, Martin (1971) *Poetry, Language, Thought*, trans. Albert Hofstadter, New York: Harper Colophon.

Johnson, Randal and Stam, Robert, eds (1982) *Brazilian Cinema*, New York: Columbia University Press.

Kershaw, Baz (1991) *The Radical in Performance: Between Brecht and Baudrillard*, New York: Routledge.

Marx, Karl and Engels, Frederick (1975) *Collected Works*, vol. 3, New York: International Publishers.

Mercer, Kobena (2005) *Cosmopolitan Modernisms*, Cambridge, Massachusetts: MIT Press.

Negt, Oscar and Kluge, Alexander (1993) *The Public Sphere and Experience*, Minneapolis: University of Minnesota Press.

Nietzsche, Friedrich (1967) *On the Genealogy of Morals: Ecce Homo*, New York: Vintage Press.

Nussbaum, Martha (2008) 'Human Rights and Human Capabilities': Twentieth Anniversary Reflections, *Harvard Human Rights Journal* 20: 21-24.

Parker, John (2007) *The Aesthetics of Antichrist: From Christian Drama to Christopher Marlowe*, Ithaca, New York: Cornell University Press.

Said, Edward (2004) *Humanism and Democratic Criticism*, New York: Columbia University Press.

Sen, Amartya (2004) 'Capabilities, Lists and Public Reason: Continuing the Conversation', *Feminist Economics* 10(3): 77-80.

Sticca, Sandro. (1967) 'The Montecassino Passion and the Origin of the Latin Passion Play', *Italica* 44(2).

Thiongo, Ngugi wa (1997) 'Enactments of Power: The Politics of Performance Space', TDR 41(3): 12.

Thiongo, Ngugi wa (1998) *'Penpoints, Gunpoints and Dreams'*: Clarendon Lectures in English Literature 1996, Oxford: Clarendon Press.

Virilio, Paul (2005) *The Accident of Art*, Cambridge, Massachusetts: MIT Press.

Young, Karl (1910) 'Observations on the Origin of the Passion Play', *PMLA* 25(2): 309-354.

Ecstasy and Pain
The ritualistic dimensions of performance practice

RINA ARYA

In this paper I want to articulate a dialogue between two counteractive narratives: the first is the Christian narrative in which the sacred is experienced in the sacrament and which presents a dichotomy between the spirit and the flesh, favouring the former. The second I will refer to as a 'post-Christian' narrative. This locates the spirit in the flesh. A sense of the sacred is experienced through acts of transgression and violation, binding the participants into a collective. In the postmodern age, following the cultural shift in the West signalled by the 'death of God', a sense of the religious can be expressed in a variety of different secular practices. The post-Christian narrative accommodates this shift of thinking. It entails a move beyond the institutions of the Church and their concomitant representations and opens up a space to rethink the narrative from *outside* the traditions and structures of its practice. It is in this space that I situate examples of performance art from Marina Abramović and Hermann Nitsch. Whether the artists in question profess to participating in this post-Christian narrative is perhaps less important than the crucial idea that they are taking part in this rethinking of the theological boundaries of contemporary thought. This meeting, or congregation, of the paradigms of theology and performance art comes out of my own research as a theologian, attempting to partake in a mapping of the interstitial spaces that lie between traditional theology and 'secular' religiousness.

Throughout my narrative I will discuss three ideas, which are central to the selected examples and are pivoted around the body: the sacred and the profane, the relationship between wounding and healing, and finally the establishment of the communal or the congregational.

THE SACRED AND THE PROFANE

In *The Elementary Forms of Religious Life*, Durkheim speaks of the sacred and the profane as being two realms or domains that are qualitatively different (Poggi 2000: 146). The realm of the profane is that of the ordinary and everyday world, while the sacred is distinct from this by virtue of being 'wholly other', 'set apart and forbidden' (Durkheim 1995 [1912]: xlvi). The reciprocal relationship between these two realms is mediated by the ritual that Rambo describes as operating as a 'bridge' between these two realms (Richardson and Bowden 1983: 509). This separation effected by the ritual is observed in daily life - a church is where we may feel the force of the holy (a related term to 'the sacred') while an abattoir is indisputably profane. The thresholds between these two realms may vary from context to context. However, what still holds is the dualistic nature of the dynamic. In the Christian narrative the sacred is experienced in what is not profane, and vice versa. However, in the post-Christian narrative the sacred is experienced through the depths of the profane, in the utter desecration of the body.

In *Rhythm 0* (1974) Abramović stood by a table and offered herself passively to spectators. A text on the wall reads, 'There are seventy two objects

 Performance Research 13(3), pp.31-40 © Taylor & Francis Ltd 2008
DOI: 10.1080/13528160902819307

on the table that can be used to me as desired - I am the object' (Warr and Jones 2000: 125). The objects varied from those that could be used as weapons - including a gun, a bullet, a saw and knives - to others that were seemingly innocuous, such as lipstick, a rose and olive oil, but which could nonetheless be used to further objectify the body. The performance lasted for six hours, and by the end all her clothes had been sliced off her body with razor blades. She had been cut, decorated, crowned with thorns and had had the loaded gun levelled against her head (125). Goldberg explains how 'this final act caused a fight between her tormentors, bringing the proceeding to an unnerving halt' (Goldberg 1988: 165). At the outset Abramović transformed herself from being the agent and author of her performance into completely submitting to the participants. Furthermore by renouncing her subjectivity, Abramović had turned herself into a thing, and the array of objects available invited the participants to explore her as a plaything. If we construe the written invitation as a contractual obligation, then it is plausible that Abramović constructed her work to function as what Linderman calls a 'limit-text', which queries boundaries that are repressed in other texts (O'Dell 1998: 3). The boundaries that are queried here are that of subjectivity-objectivity and the sacred-profane. Abramović *appears* to be subverting the function of ritual in a Christian sense because she allows participants to violate her body and desecrate her. Religious rituals serve to safeguard the sacred from the profane effects of desecration and, conversely, to protect the profane from the sacred, where the effervescent and contagious nature of the sacred may disrupt the banalities of the profane. In the series of ritual actions, the participants transgress the physical boundaries of Abramović's body. Likewise, she transgresses the psychological boundary separating subjectivity from objectivity. In an interview with McEvilley she confesses, 'I realized that the subject of my work should be the *limits* of the body. I would use performance to push my mental and physical limits beyond consciousness' (McEvilley 2000 [1983]: 15). The irony is that by desecrating these boundaries Abramović and the audience are actually consecrating the body and thereby entering into the function of the ritual wholeheartedly. This is because transgression, which involves the temporary suspension of the limit is also, paradoxically, a reinforcement of the limit.[1] Therefore, the limit or boundary is remade or recreated, and the function of the ritual reaffirmed through the action of overstepping. Given this, the rituals of desecration can be viewed from the perspective of religious ceremony. The participants enact the sacrament, which involves a consecration of the profane body.

Abramović's body is not regarded as banal and everyday but is 'set apart' by its being openly violated. By eliminating the social mores and proprieties that are tacitly observed with due regard to the treatment of an 'other', Abramović was rendering the object-body extraordinary. Parallels can be drawn with the *kenosis* of Christ, who divested or laid aside certain divine attributes, such as omniscience, when he took on human form in the Incarnation. However, Christ did not simply take the form of any man but that of the lowliest status of man and 'emptied himself, taking the form of a servant' (Philippians 2: 6-11)[2]. Metaphorically, Abramović can be viewed as abdicating her role as the author of her performance and being transformed into an abject body. In anthropological terms she could be regarded as the sacrificial victim who purges the wrongs of the community. And similarly, an analogy can be constructed with Christ who was a scapegoat for the sins of humanity.

WOUNDING AND HEALING

Taylor observes that the term 'religion' is derived from the Latin stem *leig*[3], which means 'to bind', and interprets religion - *religare* - as a binding that is a rebinding - *re-ligare* - where the rebinding is supposed to bind together what has fallen apart. Within this usage, 'religion

[1] In the performance itself, the acts of transgression serve to demarcate the boundaries. Evidence for this is provided in the action of the halting of the potentially fatal gunshot to the head. In other words, transgression does not allay the fear and potency of the taboo but actually reinforces it.

[2] See also 'The Suffering Servant' (Isaiah 53: 4-12) as another example of *kenosis*.

[3] Benveniste states that this etymology was invented by Christians and first used by Lacantius and Tertullian. Other sources (Cicero) speak of *relegere*, while *religio* remains the most constant inter-pretation (Derrida and Vattimo 1998: 36-7, 72).

therefore functions to heal the wounds, mend the tears, cover the faults, and close the fissures that rend self, society and world' (Taylor 1992: 46). Within the Christian narrative the fragments of the Crucifixion are resolved, made whole (salved) and re-bound in the Resurrection. This pattern from fragmentation (wounding) to wholeness (healing) is enacted in the rituals, which are only made possible and credible by the fragmentation of the body. Christian rituals simultaneously acknowledge fragmentation whilst also being a celebration of reconstituted wholeness. However, in this move from fragmentation to wholeness, there has been an inclination to forget the fragmentation, to forget 'Christianity's early emphasis on bodiliness and its place in the Christian liturgy' (Raitt 1993: 105). LaFleur cites the doctrines of Incarnation, Resurrection and *imago dei* and theories about the ingestion of the very body of Christ in the Mass to support his claim that Christianity is indeed 'ensconced in the semiotics of the body' (LaFleur 1998: 41, 36). Furthermore, the doctrine of the Resurrection insists on the bodily and not simply spiritual Resurrection of Christ. In the words of St Paul, 'If Christ has not been raised, then our preaching is in vain and your faith is in vain' (1 Corinthians 15: 14-15). Shaw explains the transition from the bodily to the spirit by suggesting that the concentration on the corporeality and suffering of Christ was 'as part of a greater intention to aggrandize the glory and magnificence of God' (2006: 19-20). The ultimate focus was on wholeness and salvation, and hence 'the aesthetics of the Christian sublime seeks to overcome its origins in the flesh and accomplishes this by purging the Christian sublime - which Shaw identifies as *agape* - of *eros*, thereby acknowledging the selfless love that comes from the impulse of the soul (Shaw 2006: 22-3).

While the Christian tradition locates the body in the spirit, the post-Christian tradition does the reverse, and the spirit is located in the body, the *sarx*, (the wounded flesh of Christ); there is no resolution in the sense of the body being salved and made whole. We remain only with the fragments of the body. The brutal materialism of the body in the post-Christian tradition is articulated in the performances of Hermann Nitsch[4] whose performances are characterized by an overpowering sense of the bodily and the visceral. In his combination of various Christian and pagan rites, notably the cult of *sparagmos*[5], mutilated bodies of humans and animals were strung up as if crucified. The impression of torture and brutality is further conveyed by the abundance of bodily fluids. The body is not purged of 'eros' here but is exploited in its tendency to break down and decay. In the ritualized staging of the crucifixions, the body is sundered and featured in its bloodiness and lacerations. In the Christian tradition, the sacred is experienced in the re-ligare of the fragments into a whole in the Resurrection. In the post-Christian tradition, we experience the sacred through the fragments. Through the pain and ecstasy generated by sparagmos where the animals are ripped to shreds and senses are intoxicated, the participants experience the sacred.

In the *Orgies Mysteries Theatre* (OMT) (1965) Nitsch employed a range of materials to evoke multi-sensory performances that would engage the participants on a number of levels to synaesthesic effect and induce a state of convulsion. One of the leitmotivs was the staging of the act of crucifixion. In some cases actual violence was used within the performance, such as the flaying and disembowelling of an animal carcass as well as the ongoing brutality of swinging the carcass around the room and hitting various parts of it. The human 'actors' in his dramas often had to undergo (a series of) activities that involved them adopting contortions, such as assuming the pose of the crucified, and were frequently doused in revolting liquid substances. The use of violence in these cases was mimetic (and not actual) and was meant to mirror and replicate the brutality enacted on Christ on the Cross and conveys the solemn moment at which sentient flesh is turned

4 Although it credible to claim that the collective work of the Vienna Actionists, who included Nitsch, were oriented towards the body in its fragments, I have chosen to isolate Nitsch from this group because his work is directly inspired by the symbolism of Christianity and thus has greater pertinence.

5 Paglia defines *sparagmos* as the violent principle of Dionysius, which means 'a rending, tearing, mangling' and, secondly, 'a convulsion, spasm' (Paglia 2001: 95).

into meat, into a thing. In *Action 1* (1962) a young man, dressed in a white shirt was tied to the wall of a room as if crucified and blood was poured and squirted over the bound man's head from small containers and a colonic irrigator (Green 1999: 135). Although the blood is not issued from an actual wound, the discomfort caused by the stream that runs down his face and clothing conjures up the impression that he is wounded. Jones comments on how, 'Nitsch aimed an ecstatic redemption through the powerful emotional experience of the physical contact with blood and by assuming the role of Christ' (Warr and Jones 2000: 92).

When questioned about the effects that his performances had on him, Nitsch responded that 'it frightens and fascinates me at once' (Montano 2000: 399). His phrase bears marked similarities to the phrase used by the theologian Rudolf Otto to designate the feeling on encountering the holy, the *mysterium tremendum et fascinans,* which is to be distinguished from the 'profane, non-religious mood of everyday experience' (Otto 1958: 19, 12). Nitsch seeks to express the experience of the sacred. In discussions of his work Nitsch uses religious language to support his intentions and motivations. The animal carcass or human body stands in for the Christ figure, and Nitsch through his orchestration dramatizes the crucifixion in all its religious horror. In *The Blood Organ* (1962) he extends his metaphoric range by actually taking on the role of Christ. Nitsch plays the sacrificial victim or scapegoat who carries the burden of responsibility so as to absolve others of the bleakness of humanity and states how:

> I take it upon myself all that appears negative, unsavoury, perverse and obscene, the lust and the resulting sacrificial hysteria, in order to spare YOU the defilement and shame entailed by the descent into the extreme. (Green 1999: 132)

Nitsch's blasphemous appeal was intended to be provocative. He wanted to revive the significance of the Crucifixion for a contemporary audience and felt that the bloodiness and violence had been sanitized and made banal over the centuries. By provoking a direct, visceral and sensory experience, the participants would be able to release powerful pent-up emotions that had been hitherto repressed. He was advocating a psychoanalytical approach that would encourage the release and expression of animal instincts. This cathartic outpouring can be described as a form of healing but not within the framework of the Christian tradition, where healing is equivalent to wholeness. In the post-Christian example of Nitsch, healing occurs by meditating on the fragments, on the wounded nature of the body. And through catharsis the sanctity of the symbol of the Crucifixion is restored. Enlightenment and the overarching sense of communion would occur only through extreme measures. Nitsch describes how he had to resort to 'histrionic means', which were harnessed to gain access to the profoundest and holiest symbols through blasphemy and desecration (Green 1999: 132).

Through the employment of violence or gore Nitsch succeeded in expressing the original sense of crucifixion. He disfigures the symbol of the Crucifixion by dislocating its position in the Christian narrative and re-locating or positioning it within what we would now understand as a Girardian context (of the violence of the sacred), where it is returned to its primitive origin, as a symbol of punishment or agony. In pre-Christian times, the cross was regarded as a technical device, which referred to the mode by which people were put to death. It therefore referred to the instrument that effected the torture. In Roman times crucifixion was widely used as a form of capital punishment reserved for criminals. Martin Hengel describes it in the following way: 'far from being a dispassionate execution of justice ... the crucifixion satisfied the primitive lust for revenge and the sadistic cruelty of individual rulers and the masses' (Moore 1994: 96). By revealing what lies behind the sanitized veneer of the symbol, the participants experience the 'real presence'[6] of violence, which is arguably

[6] The 'Real Presence' is a phrase used in Eucharistic theology to express the actual rather than figurative presence of the Body and Blood in the Sacrament.

more religious because of its proximity to the Passion of Christ. The theological analogue of this moment is the cry of desertion at the ninth hour, 'My God, my God why hast thou forsaken me?' (Matthew 27: 46-7). Nitsch's intentionally provocative stance affected the way he was received by critics and public alike. Green discusses the various portrayals of Nitsch that range from 'a parodist of religion' to 'more pious than the Pope' (Green 1999: 13). These seemingly contradictory depictions convey both aspects of Nitsch's perspective. He was criticizing the bland and banal constructions of the Crucifixion and in that respect was anti-Christian. And yet by deconstructing the symbol and exposing the unremitting brutality that lies at the crux of the Christian religion, he was reviving the gravitas and significance of the sacrifice. He does not present a salvific narrative, and we remain with the bodily fragments and mental sparagmos, which is generated by the disorder of senses. Other designations that are consistent with his intentions include: he was 'a modern day Grünewald', 'a kind of latter-day St Francis, who has brought home the central mystery of Christianity, the bloody sacrificial death of the Redeemer' (Green 1999: 13).

THE WOUNDED COMMUNITY

> If human beings had kept their own integrity and hadn't sinned, God on one hand and human beings on the other would have persevered in their respective isolation. A night of death wherein Creator and creatures bled together and lacerated each other and on all sides, were [sic] challenged at the extreme limits of shame: that is what was required for their communion.
>
> (Bataille 2000 [1945]: 18)

In the above excerpt, from *On Nietzsche*, Bataille discusses how communion was achieved through the wound that was created between God and humankind, as instantiated in the Crucifixion. He is emphatic that this sense of community and intimacy 'cannot proceed from one full and intact individual to another. It required individuals whose separate existence in themselves is *risked*' (Bataille 2000 [1945]: 19) and is in a state of *ek-stasis*. Abramović and Nitsch conform to a similar model, whereby the wounded body (of the artist) generates the feeling of community. The wounded body operates as a channel, which shatters the boundaries separating the self from the other, and the identity of the participants is consolidated into a collective whole. They also implicate the audience-cum-participants in the doing/making so that the tension and anxiety created through the enactments of certain rituals, inviting others to harm (in the case of Abramović) and sacrifice (in Nitsch's work) serve to psychologically wound the participants. The effects of wounding encourage a sense of fellow-feeling and solidarity, which engenders a sense of 'community'. This understanding of community is closer to the original conception of the Church, which referred to the body of believers rather than to the church building. In the post-Christian rethinking of theological boundaries, the church as building is then supplanted by a new type of community and membership.

This transformation of identity from individuality to community is effected by abreaction. Adapted from Breuer's and Freud's study of 1893 in 'On the Psychical Mechanism of Hysterical Phenomena: Preliminary Communication', 'abreaction' is one of Nitsch's central concepts. It refers to 'a release or discharge of emotional energy following the recollection of a painful memory that has been repressed' (Colman 2003: 3). This release of blocked energies occurs during moments of physical and physiological violation. In abreaction 'not only are the individual's blockages released, but the floodgates are opened to the immeasurable' (Green 1999: 261). Nitsch compiles a series of abreaction rites used in *OMT*, which involve activities that cause the derangement of the senses. For example, 'raw meat will be torn apart', 'ecstasies are created by producing the loudest possible noise' and 'the power of speech is reduced to liberating cries of lust'. Nitsch explains that the choice of rite was determined according to 'a scale of stimuli which can produce a meditative and prodigiously sensual feeling in the individual' (Green 1999: 134). In *Rhythm O*, the participants experience a sense of Dionysian excess and the carnivalesque as they subject Abramović to a range of activities. The physical violence is transformed into psychological violence, which causes disarray in the collective body of the participants. The transition and transformation of identity from individualism to the suspension of individualism in lieu of a collective identity can be viewed from the perspective of Gluckman's theories of ritual, which discusses how ritual is a mechanism for constantly re-creating and not simply reaffirming the unity of the group (Bell 1997: 39). Another switch that occurs is the transition from objectivity to subjectivity. In a subversive turn, the artist assumes a submissive role and the participant becomes the author of making. In *Rhythm O*, Abramović presents a situation in which the viewer 'implicates him/herself in the potentially aggressive act of unveiling and marring her passive body'[7]. The participants are bound together through their collective actions or sentiments. They are bound by their experience of the sacred horror of the desecration of the body into the status of a thing.

[7] This description is used with reference to the role of Yoko Ono in *Cut Piece* (see Warr and Jones 2000: 74), but I have purposefully widened its application here.

The formation of communal bonds and the collective identity generated in these performances reconfigures the expectations of the spectators as well as the roles that are bestowed upon the audience. My deliberate use of 'participants' rather than 'audience' was to convey the interaction and intersubjectivity that is engendered through the performance itself. The reciprocal relation that exists between 'artist' and 'participant' perpetuates this sense of community and is created strategically through contractual exchange. In *Rhythm O*, Abramović posts a written invitation to the participants explaining what they are required to do. Nitsch's approach is more forthright; indeed in his *OMT*, nobody is allowed to enter his performance site merely to observe - participation becomes obligatory. The hierarchical divide separating artist from consumer is eliminated in a Brechtian turn to impose an uncomfortable and self-conscious state on the audience in an attempt to reduce the gap between the two (Goldberg 1988: 162).

The communal identity created in the performances also has important anthropological and social functions. The mutilation of Abramović can be paralleled to the function of ritual sacrifice in paganism and, in Christian terms, to the sacrifice of Christ on the Cross. In all cases, the body operates as the meeting point between opposites, between self and the other, the sacred and the profane, and life and death. The sacrifice of the body purges the community of violence, thus restoring harmony and equilibrium. In the performances, no one is killed (one of Nitsch's preconditions is the use of only dead animals). However, what does happen in this meeting of opposites is that our assumptions as viewers and our conditioned responses to the body are quashed. The Cartesian subject of Modernism is dislocated, as is the gender-oriented axis of perception. By testing the boundaries of the body, whether through the physical actions that we inflict on the artists' bodies or by our psychological perceptions of what is happening, we experience knowledge *through* the senses. In his formulation of the 'Theatre of Cruelty', Artaud states that the masses 'think with their senses first and foremost' and that an experience of mass theatre should explore 'the limits of our nervous sensibility' (2000 [1933]: 216). Through the exploration of nerve-exposed sensations, the body is undone, is taken apart, which is simultaneously a form of embodying: we learn what it feels like to be embodied. The objects are also transformed through action. In Rhythm O, the critic Stooss states that during the performance the objects were supposedly transformed into art objects in their own right (Abramović and Abramović 1998: 14); they are transmogrified.

THE 'REAL PRESENCE' OF THE BODY

In the performances of Abramović and Nitsch, the participants gain through a sense of loss. They gain an understanding of embodiment by experiencing a sense of the disintegration of language that concurs with the vertiginous experiences and the accompanying loss of a sense of self. This tendency mirrors the practice of the mystics, who through their resolute denial of the drives of the body ironically place it at the forefront of their quest. Vergine states that 'mysticism is first of all a physical experience: a source of fluids, of bloods, of humours, of various waters that flow, coagulate, and again grow liquid' (2000: 291). The mystics regarded the moment of the loss of self as a moment of self-discovery, and this simultaneous shift can be seen within the context of performance art. It is plausible that the artists employed their bodies or the bodies of others and inflicted extreme experiences on these bodies in order to create an alteration of perception. If that is the case, then the performances are not about the body at all but rather are about the mind. They probe us to interrogate our preconditioned responses to the dualisms that we set up between subject and object, pain and pleasure, and viewing and participation. Whether their intentions are to provoke intellectual questions is not the central issue, however. Rather, what seems undeniable is their response to a technological society where corporeality has become sanitized and marginalized. Sobchack makes an incisive distinction between the viewing of the body and the feeling in the body. Commenting on the effects of globalization and mass media, Sobchack states that:

> To say we've lost touch with our bodies is not to say we've lost sight of them. Indeed, there seems to be an inverse ratio between seeing our bodies and feeling them: the more aware we are of ourselves as the 'cultural artifacts', 'symbolic fragments' and 'made things' that are images, the less we seem to sense the intentional complexity and richness of the corporeal existence that substantiates them.
> (Warr and Jones 2000: 41)

In the performances we are not permitted simply to view the bodies, but we experience them in their sentient states and even initiate these states of wounding. We see behind the 'cultural artifacts' and 'symbolic fragments', behind the body as representation and experience the re-presentation of the body. We see behind the symbol to the 'real presence' of the sacrament. This is experienced in the intoxicated states that the performances induce in the participants. In his merger of the myth of the Crucifixion with the slaughtering of animals in an abattoir, Nitsch restages the Crucifixion in its original locale, in Golgotha. In her video installation *Spirit House* (1997), which bears structural similarities to the Passion, Abramović presents five separate video sequences, called 'stations', one of which involves her flagellating herself with a whip. Its pertinence is exacerbated by the first presentation of the installation at Caldas da Rainha, Portugal, in 1997 in a former communal slaughterhouse (Abramović and Abramović 1998: 13). By placing what I interpret as Abramović's 'stations of the Cross' in a slaughterhouse, she is parallelling the experience of the Crucifixion with the slaughtering of an animal thereby enhancing the brutality of the suffering of Christ. This questions the legitimacy of the killing in the first place—how can such a treatment be sanctioned? The only justification lies in the fulfilment of the Christian narrative, where eternal life follows death and the sacrifice of the Son confirms God's love for humankind. If we remove this teleology, we are left with the bloody murder on the Cross. Moore conveys the reprehensibility of the act in a very direct and immediate manner:

> My own father too was a butcher, and a lover of lamb with mint sauce. As a child, the inner geographical boundaries of my world extended from the massive granite bulk of the Redemptorist church squatting at one end of our street to the butcher's shop guarding the other end. Redemption, expiation, sacrifice, slaughter ... There was no city abattoir in Limerick in those days; each butcher did his own slaughtering. I recall the hooks, the knives, the

cleavers; the terror in the eyes of the victim; my own fear that was afraid to show; the crude stun-gun slick with grease; the stunned victim collapsing to its knees; the slitting of the throat, the filling of the basins with blood; the skinning and evisceration of the carcass; the wooden barrels overflowing with entrails; the crimson floor littered with hooves.

I also recall a Good Friday sermon by a Redemptorist preacher that recounted at remarkable length the atrocious agony felt by our sensitive Saviour as the spikes were driven through his wrists and feet. Crucifixion, crucifixiation, crucasphyxiation ... Strange to say, it was this sombre recital, and not the other spectacle, that finally caused me to faint. Helped outside by my father, I vomited gratefully on the steps of the church. (Moore 1996: 4)

Moore's parallel conveys the similarities between the activities that are carried out in the butcher's shop and the Crucifixion of Christ. However the former domain is more defensible than the signification of events being carried out in church because it does not misconstrue its practice. However, in church the 'atrocious agony felt by our sensitive Saviour' is used as a pretext for belief in the Christian narrative. The sermon recounts a brutality that is teleological. And it is this chain of events that causes a turbulence of sensation for Moore. Ironically, he moves from the symbolic event to the physical reality - from 'crucifixion' to 'crucifixiation' to 'crucasphyxiation' to delegitimize the signification. The associative wordplay parodies the justification and conveys the accumulated sense of nausea that finally causes Moore to seek relief on the steps of the church. In this post-Christian tradition there is no atonement, and we are punished for our tendency to anaesthetize the violence at the heart of the Cross. The proximity of our fate to that of the animals in the butcher's shop cannot be overstated.

CONCLUSION

The essay sets up a dialogue between Christianity and post-Christian thinking. In the commentaries on the examples of performance art, I demonstrated that although these examples do not support or uphold the Christian narrative of salvation where fragmentation is resolved through *re-ligare* in the Resurrection, they are arguably more religious because of the experience of the sacred that these works give rise to in the crossing from one boundary to another - from subject to object, from life to death, which can be described as sacred. The sacred is experienced in the excess of this transgression, which generates a feeling of 'collective effervescence' (to use a Durkheimian phrase) in the participants. And as with the nature of the sacred, the experience cannot be recollected in representation. The participants experience the contagion and violence of the sacred, which is expressed in the depths of the profane. Nitsch declares how the passage to the sacred is reached through the profane: 'histrionic means will be harnessed to gain access to the profoundest and holiest symbols through blasphemy and desecration. Blasphemous provocation is tantamount to worship' (Green 1999: 132).

In the space opened up by the 'death of God', the sacred is not experienced in transcendence but in the resolute immanence and brute materialism of the body. I suggest that the works demonstrate an a-theology of incarnation. The term, 'a-theology', coined by Mark C. Taylor, was employed to accommodate such practices, which cannot be described as theological per se but which rethink the margins of theological thinking. In the interstitial space between theism and atheism, 'we find the possibility of refiguring the polarities and oppositions that structure traditional religious thought' (Taylor 1992: 4). The performances enact, perform and instantiate the intertwining of self and other (Jones 1998: 38), and we experience what it feels like to be embodied, which in Christian terms is deeply incarnational. They articulate a counter-narrative to the bodies of the powerless created through the institutionalized power of the church and invite the possibility of narratives of spiritual empowerment.[8] In a reverse of the

[8] The examples can also be viewed from the perspective of liberation theology.

Augustinian model in *Confessions* (397) where continence is the route to the spiritual life, here we have a move towards the wounded body as the practice of religiosity. Earlier I listed a series of designations to describe Nitsch, which included the opinion that he is 'a modern day Grünewald'. I wish to develop this a step further by suggesting that while Grünewald articulated theology in his painted depictions, the performance artists take this further, and the body becomes the holy text, where the performance of the text can be described as 'an aesthetic liturgy' (Vergine 2000: 117).

Christianity is often kept out of the performance arena, but in this meeting of theology and performance practice we experience a synergy, which rehabilitates both practices. Theological thinking is revised, updated and moved from its doctrinal and institutionalized focus to a post-Christian understanding of the significance of its rituals and practices. The examples of performance practice paradoxically revitalize the religiosity of Christianity because they take us to the beginnings of theology, with the violence of the sacred. They also make us reconsider the teleological legitimacy of the Crucifixion. The 'real presence' of the performances operates as a contrast to the sanitized representations of the institutions of Christianity. The congregational meeting informs performance practice by demonstrating that in contemporary society theology occurs outside the boundaries of the Church and in a variety of different contexts. And the artists operate as commentators who grapple with critical questions in contemporary society. In this technological age of consumerism, such a 'poetics of agony' enables us to move beyond the representation to a more embodied understanding of humanity. Thus the implications are not merely aesthetic but also ethical. This meeting of theology and performance art is mutually illuminating.

REFERENCES

Abramović, M. and Abramović, V. eds. (1998) *Marina Abramović: Artist Body: Performances 1969-1998*, with texts by Velimir Abramović, Jan Avgikos, Chrissie Iles, Thomas McEvilley, Hans Ulrich Obrist, Bojana Pejić, Tony Stooss, Thomas Wulffen, Milano: Charta.

Artaud, A. (2000 [1933]) 'Theatre and Cruelty', trans. V. Corti, in Tracey Warr and Amelia Jones (eds) *The Artist's Body*, London: Phaidon, p. 216.

Bataille, G. (2000 [1945]) *On Nietzsche*, trans. B. Boone, London: The Athlone Press.

Breuer, J. and Freud, S. (1893) 'On The Psychical Mechanism of Hysterical Phenomena' in *The Standard Edition of the Complete Works of Sigmund Freud, Volume II (1893-1895)*: Studies on Hysteria, 1-17.

Bell, C. (1997) *Ritual: Perspectives and Dimensions*, New York: Oxford University Press.

Bell, C. (1998) 'Performance' in M. C. Taylor (ed.) *Critical Terms for Religious Studies*, Chicago and London: University of Chicago Press, pp. 205-24.

Burden, C. (1984 [1975]) 'Through the Night Softly' in Gregory Battcock and Robert Nickas (eds) *The Art of Performance: A Critical Anthology*, New York: E. P. Dutton, pp. 222-39.

Colman, A. M. ed. (2003) *A Dictionary of Psychology*, New York: Oxford University Press.

Derrida, J. and Vattimo, G. eds (1998) *Religion*, Cambridge: Polity Press.

Durkheim, E. (1995 [1912]) *The Elementary Forms of Religious Life*, trans. K. Fields, New York: The Free Press.

Gorsen, P. (1984 [1979]) 'The Return of Existentialism' in Gregory Battcock and Robert Nickas (eds) *The Art of Performance: A Critical Anthology*, New York: E. P. Dutton, pp. 135-41.

Goldberg, R. (1988) *Performance Art: From Futurism to the Present Day*, London: Thames and Hudson.

Green, M. ed. (1999) *Brus, Muel, Nitsch, Schwarzkogler: Writings of the Vienna Actionists*, trans. M. Green, London: Atlas.

Howell, A. (2000) *The Analysis of Performance Art: A Guide to its Theory and Practice*, Amsterdam: Harwood Academic.

Jones, A. (1998) *Body Art / Performing the Subject*, Minneapolis: University of Minnesota Press.

LaFleur, W. R. (1998) 'Body' in M. C. Taylor (ed.) *Critical Terms for Religious Studies*, Chicago and London: The University of Chicago, pp. 36-54.

McEvilley, T. (2000 [1983]) 'Art in the Dark' in Tracey Warr and Amelia Jones (eds) *The Artist's Body*, London: Phaidon, pp. 222-7.

Miles, M. R. (1993) 'Desire and Delight: A New Reading of Augustine's *Confessions*' in M. A. Tilley and S. A. Ross (eds) *Broken and Whole: Essays on Religion and the Body*, Lanham, Maryland: University Press of America, pp. 3-16.

Montano, L. M. ed (2000) *Performance Artists Talking in the Eighties*, Berkeley and Los Angeles: University of California Press.

Moore, S. D. (1994) *Post Structuralism and the New Testament: Derrida and Foucault at the Foot of the Cross*, Minneapolis: Augsburg Fortress.

Moore, S. D. (1996) *God's Gym*, New York: Routledge.

O'Dell, K. (1998) *Contract with the Skin: Masochism, Performance Art and the 1970s*, Minneapolis: University of Minnesota Press.

Otto, R. (1958) *The Idea of the Holy: An Inquiry into the Non-Rational Factor in the Idea of the Divine and its Relation to the Rational*, trans. J. W. Harvey, Oxford: Oxford University Press.

Paglia, C. (2001) *Sexual Personae: Art and Decadence from Nerfertiti to Emily Dickinson*, New Haven and London: Yale University Press.

Pluchart, F. (1984 [1978]) 'Risk as the Practice of Thought' in Gregory Battcock and Robert Nickas (eds) *The Art of Performance: A Critical Anthology*, New York: E. P. Dutton, pp. 125-34.

Poggi, G. (2000) *Durkheim*, Oxford and New York: Oxford University Press.

Raitt, J. (1993) 'Christianity, Inc.' in M. A. Tilley and S. A. Ross (eds) *Broken and Whole: Essays on Religion and the Body*, Lanham, Maryland: University Press of America.

Richardson, A. and Bowden, J., eds (1983) *A New Dictionary of Christian Theology*, London: SCM.

Scarry, E. (1985) *The Body in Pain*, New York: Oxford University Press.

Shaw, P. (2006) *The Sublime*, Abingdon: Routledge.

Stiles, K. (2000) 'Quicksilver and Revelations' in L. M. Montano (ed.) *Performance Artists Talking in the Eighties*, Berkeley and Los Angeles: University of California Press, pp. 473-92.

Sylvester, D. (1987) *Interviews with Francis Bacon*, London: Thames and Hudson.

Taylor, M. C. (1992) *Disfiguring: Art, Architecture, Religion*, Chicago: University of Chicago Press.

Vergine, L. (2000) *Body Art and Performance*, trans. H. Martin, Milan: Skira.

Warr, T. and A. Jones eds (2000) *The Artist's Body*, London: Phaidon Press.

‘Secular Sacredness’ in the Ritual Theatre of Nicolás Núñez

DEBORAH MIDDLETON

> Núñez has forged a theatre that attempts to provide an active arena for those individuals who seek personal individuation, communitas, or the experience of the sacred dimension in their everyday lives. The TRW carves out a sacred space in the context of an urban cultural institution.
>
> (Middleton 2001: 43)

The work of Nicolás Núñez and his collaborators at the *Taller de Investigación Teatral* (Theatre Research Workshop / TRW) in Mexico City offers an activity that responds to both ritual and theatrical imperatives and that integrates religious sources as transferable psycho-physical practices. Núñez calls this an ‘anthropocosmic theatre’, a theatre of ‘high risk’ for the purposes of personal transformation (Núñez 2007). He locates the work within a pan-cultural ‘secular sacredness’, which seeks to reconnect theatre with its archaic ritual sources.

I first began researching Núñez’s work in 1993, after encountering his training ‘dynamics’ at a conference on Performance, Ritual and Shamanism at the Centre for Performance Research in Wales. The article ‘At Play in the Cosmos’, quoted above, was the result of eight years of engagement in Núñez’s work, in residencies that I organized in the UK and during research trips to Mexico. In that period, and since, I have had open access to workshops, daily training, rehearsals, planning meetings and performances. There has been ample opportunity for formal interviews and long informal conversations with Núñez and his collaborators. My role in the work has been that of participant-observer, attempting to maintain that careful line between immersion in experience and the critical distance of the scholarly engagement.

In ‘At Play’ I was interested in framing Núñez’s practice as a ritual activity, elucidated by recourse to ritual structures and concepts and defined by an imperative towards sacrality. In this paper I have explored bases for an understanding of the psycho-physical causes and effects of sacred experience within the practice. This paper arises out of my deepened experience; I am now in the second half of my second decade in the work, a chapter characterized by, among other things, periods of immersion in the dynamic *Citlalmina* (since in 2000, Núñez authorized me to practice the form independently). This writing is informed by my phenomenological experience of the work over time, but for reasons of academic distance, I have preferred to draw directly on the experience of other participants as illustration and evidence.

Núñez’s project, and indeed my own engagement with it, are essentially intercultural explorations. While I acknowledge the ethical questions surrounding that aspect of the work, the focus in this article will be on the ways in which such a theatre functions as ‘a sacred space’. By identifying the culturally syncretic context in which Núñez lives and works, I hope to provide a perspective which both normalizes and legitimizes the intercultural aspects of his project.

Furthermore, I will approach Núñez’s religious sources, as he does, as performance practices

Performance Research 13(3), pp.41-54 © Taylor & Francis Ltd 2008
DOI: 10.1080/13528160902819315

that embody pre-cultural frameworks that may be uncovered to produce sets of transferable principles and psycho-physical technologies.

To this end, I will explore religious experience per se through a consideration of the consciousness-states and structures involved. Through ascertaining the bodily correlates and mental factors that are found to be instrumental in producing religious - or 'altered' - states (and structures) of consciousness, I will propose a model for producing - or increasing access to - culturally 'unformed' psycho-physical experience. Against this framework of ideas, I will map the phenomenological experiences of participants in Núñez's practice, asking to what extent they might be termed 'religious' or 'spiritual' experiences and under that specific conditions this aspect of the work flourishes.

A MESTIZO THEATRE

Núñez's project, since founding the TRW under the auspices of the National Autonomous University in Mexico in 1975, has been to create a theatre for contemporary Mexicans - a *mestizo* theatre, which draws on indigenous as well as European cultural heritages and which reflects the ancient anthropocosmic impulses of the Toltec-Mayan people. Indeed, Núñez calls his an 'Anthropocosmic Theatre'. It takes the form of a series of participatory 'dynamics', used independently as a kind of training and embedded in theatrical devices and structures as a ritual theatre.[1]

The legitimate theatre in Mexico has long been dominated by European influences. At the same time, pre-Colombian ritual performances have survived among the many indigenous peoples of Mexico, often doing so through processes of acculturation. The TRW's imperative has been to create a Mexican theatre which reflects the cultural background of the majority of the population (*mestizo*) by combining European and indigenous Mexican cultural forms. To this end Núñez has intensively explored such ritual sources as the Nahuatl conchero dance tradition, Huichol shamanic (*mara'akame*) practices, psycho-physical energetic positions derived from phenomenological research into the ancient practices suggested by archaeological artefacts, and imagery derived from the rich pre-Colombian mythologies, notably that of the culture hero, *Quetzalcoatl*.

This is a complex set of cultural sources; Núñez navigates a terrain characterized by religious and cultural syncretism. Indeed, Timothy Light, a scholar of Comparative Religion, considers that Mexico's national symbol, The Virgin of Guadalupe, is 'as epitomizing an example of syncretism as can be found' (2004: 334). This national symbol, who is both Mary and the Nahuatl fertility goddess or Nature symbol, *Tonantzin*, is appropriate in a country where pre-Colombian and colonial religious and cultural forms are intricately interwoven. Widespread cultural syncretism is a lived reality for contemporary Mexicans. The conchero dance tradition, so central to the TRW experience, is, like *Maria-Tonantzin*,[2] a tight knit of pre-Hispanic Nahuatl performance and Spanish Catholicism.

It is beyond the scope of this essay to explore the politics of colonial hegemonies, cultural adaptations and resistances. I would, however, like to identify religious syncretism as a context for the work of the TRW. For, as we shall see, Núñez and his collaborators have embraced that tolerance for synthesis and extended it to religious practices from beyond their own mixed cultural heritage.

The TRW might be described as a kind of theatre/anthropological investigation, in Barba's sense (1991), in which attention is focused on the fundamental performative technologies within the sources, eliminating all contexts but the psycho-physical.[3] This essay seeks to understand to what extent religious practices can be stripped down in this way, and represented as theatrical devices which are, nevertheless, designed to cultivate a wholly secular spiritual experience. I have previously claimed that 'Núñez is particularly careful not to invest the actions and experiences of the dynamics with any overt

[1] Here, we will be concerned mainly with the activities that fall within the training sphere of Núñez's work, the 'dynamics', and will not address the further problem of whether individuals encountering the work only through a single experience of a ritual theatre production might also have heightened access to what might be identified as the 'religious' form of an altered state of consciousness.

[2] In 1998, five members of the TRW, working under the direction of long-term member Ana Luisa Solís Gil, created the dynamic 'Tonantzin/ Maria'.

[3] For a description of two of Núñez's dynamics in these terms, see Middleton (2001).

ideology. The TRW professes no particular belief system' (Middleton 2001: 43). This has been challenged by Antonio Prieto Stambaugh, who responded,

> While Workshop members indeed avoid imposing a religion or an ideology on participants, there's a particular corpus of beliefs underlying their work. The TRW is intimately associated with a movement of spiritual nationalism known as Nueva Mexicanidad (New Mexicanity), which spans everything from Carlos Castenada to *conchero* dancing ... One of the main leaders of this movement is Antonio Velasco Piña, who is cited in the TRW's playbills as a key advisor. (2001: 8)

It is true that the TRW engage with both Velasco Piña and the practices mentioned. While I am not aware of Núñez himself having a strong involvement with that movement, some members of Núñez's circle of close collaborators are deeply involved in conchero dancing and other forms associated with the Nuevo Mexicanidad movement. And, as we shall see later, Prieto Stambaugh identifies the work of the TRW as a collective pursuit more than one authored by Núñez himself. Thus he makes a strong connection between the personal orientation of members of the group and the group's practices. In my extensive experience of the TRW's work since 1993, however, I would argue that Núñez's collaborators play a significantly different, though absolutely crucial, role in the development of the dynamics. Núñez, himself, talking about his long-term partner in the work, Helena Guardia, described the relationship thus:

> I direct the *Taller* [Workshop] and its actions. Helena supports and enriches them. I'm responsible for the outside and Helena is a kind of internal catalyser ... nothing the *Taller* has arrived at could have been achieved without her. (Núñez 1997)

Similarly, the phenomenological reports of collaborators of being inside the work have fed Núñez's research into the precise psycho-physical forms, rhythms and meditations that have the potential to create numinous experiences. We will explore below the exact nature of the relationship between belief systems brought to a practice and the fundamental nature and efficacy of the practice itself; between religious experiences that are 'formed' by culture and those that are pre-cultural and 'unformed'.

Núñez appropriates from religious and spiritual contexts only psycho-physical practices and anthropocosmic intentions; in the absence of any further instructions or ideological contexts, he offers these to participants as 'theatrical games that will allow ordinary participants to access the realm of the sacred' (Middleton 2001: 47). If Núñez's collaborators and participants experience the work through personal cultural constructs and belief systems, it is Núñez's task to identify the core physiological and cognitive processes that will facilitate the core experiential state itself.

• **Nicolás Núñez with Ana-Luisa Solís and Patricia Torres (L to R) dancing *Citlalmina*.**
Photographer: Steve Forrest.

In this way, Núñez has explored and integrated ritual and performance sources from beyond Mexico, notably the Tibetan Buddhist monastic 'Black Hat' dance, which forms one half of the dynamic *Citlalmina*, alongside a *Nahuatl* conchero, or shell dance. Núñez uses these sources with the express permission of the Dalai Lama and the leading religious authority from the Conchero tradition (Middleton 2001: 54–6). In both cases, Núñez learned the forms as they are taught within their respective religious contexts. In creating *Citlalmina* from these sources, he respected the 'body alphabets', structures and rhythms exactly but abandoned costumes and other artefacts where they did not play an essential role in the psycho-physical experience of the form. For example, the vibratory qualities of each dance are preserved; the Nahuatl shell, seed anklets (*ayoyotes*) and rattles are used in *Citlalmina*, the bone trumpets of the Black hat dance are replaced by a vibratory vocalization (the sacred syllable *Hu*). *Citlalmina* is a cornerstone of the TRW's practice. Performed weekly throughout the year in Mexico City and open to participants who range from students of acting to ordinary Mexicans wishing to engage with their cultural heritages, it represents a major vehicle for Núñez's participants to engage in a secular pursuit of sacred experience.

SECULAR SACREDNESS

Before addressing Núñez's practice in any further detail, then, we need to understand the parameters of the term 'secular sacredness'. What is it, first of all, that we refer to when we say 'the sacred'? While recourse to the presence or power of a God is a usual and convenient shortcut, there is something necessarily circuitous in this, since our experience of that which we call 'God' *is* or is *via* that which we call the sacred. Within the field of Religious Studies, there are a number of definitions of the sacred: that which is 'wholly other' (Durkheim 1995 [1912]); 'absolute' or 'ultimate' or of the highest (transcendent) value (Eliade 1957); that which inspires 'awe', the *mysterium tremendum*, the 'numinous' (Otto 1923 [1917]); that which conveys a profound sense of the connectedness of an ultimate reality, believed to underlie daily reality (Huxley 1944). These are not exclusive categories but rather differences of emphasis and approach.

For Durkheim, Eliade, Otto, Huxley, and for William James (1960 [1902]), the 'sacred' does not, by necessity, imply a deity, nor is it definitively bound up with the domain of religion (which I take, here, to imply both divinity and organized belief system). Let us, then, look more closely at the key characteristics of the sacred, as experienced and articulated across cultures.

The 'sacred' is inferred:

by **PHENOMENAL EXPERIENCE** suggests →	by **CONCEPTUALIZATION** in contrast to the 'known'
of **awe, the numinous, *mysterium tremendum*** ↓	described as **other, absolute, *ultimate, highest*** ↓
conveys experience of **profound connectedness** ↓ cognized as →	characterized by **unity of all things** ↓
equates to **'being'** in contact with →	believed to be **underlying reality**

Thus, our sense of the sacred seems to be forged on phenomenal experiences that are of an order 'wholly other' to those of daily - mundane - existence, from which we infer categories relative to that of mundane existence. Those categories are either articulated in qualitative terms, or they imply realities that are, by definition, beyond the range of our conceptual scope. And yet, despite this reaching into a conceptual void, the world's religious, spiritual and transcendentalist literatures represent a surprising coherence. Arthur Deikman, a psychiatrist researching in the field of consciousness and mysticism, states,

> Profound connection is what the word 'spiritual' properly refers to ... At its most basic, the spiritual is the experience of the connectedness that underlies reality. (2000: 84)

The spiritual quest is one that attempts to draw the subject into relation with a sacred realm, in which we experience a connectedness to an underlying reality that is itself one of profound connectedness between all things.

Thus, the 'sacred' represents for us a ground of being that the world of daily, subjective existence cannot provide. For Eliade,

> [r]eligious man's desire to live *in the sacred* is in fact equivalent to his desire to take up his abode in objective reality, not to let himself be paralyzed by the never-ceasing relativity of purely subjective experiences, to live in a real and effective world, and not in an illusion. (1957: 28)

As we shall see later, the sacred as objective reality may be understood as an experience brought about by a perceptual shift from a daily mode to one that is not bound by instrumental object-consciousness. From our isolated subjectivity behind veils of perceptual and conceptual consciousness, we are understood to break through to a reality that our daily modes of being can never reach. In that reality, the mystics tell us, the illusion of separateness is destroyed.

For Eliade, 'the sacred is equivalent to a *power*, and, in the last analysis, to *reality*. The sacred is saturated with *being* ... religious man deeply desires to be' (Eliade 1957: 12-13). Thus, contact with the sacred infuses us with an ontological certainty and energy that is not supported in the mundane world. Our phenomenological being-state is intricately connected to our engagement with an objective Universe. Contact with the sacred releases us from the isolation of a world in which we are doomed to perceive only 'fragments of a shattered universe' (24). For Eliade, the experience of the sacred 'founds the world' and thereby creates 'cosmos' (23, 29).

As we know, Núñez identifies his theatre as 'anthropocosmic' - a theatre of the human in the cosmos. Guided by pre-Hispanic mythological sources, Núñez sees a theatre for Mexico as necessarily embracing the ancient practices believed to assist people in 'aligning themselves with the cosmos'. Núñez's professed aspirations are to provide participants with experiences in which they may achieve a sense of connectedness to an underlying reality. Psycho-physical experiences, supported by mythological imagery, are designed to act as imaginal 'thresholds' or 'portals' to an experience of reality that will, in Eliade's terms, 'sacralize the cosmos' (Eliade 1957: 17) and increase access to 'being'. For Núñez, the Actor is a sacred animal, alongside the bull, the deer, etc. The sacrality of the Actor resides within the ability to access heightened states of being in which perception alters, such that we feel that we bypass cognitive conceptualizing and directly encounter the cosmos. Núñez tells us

> we have to make clear that the shaman or the actor is someone who, at will, can go into an altered state of consciousness, go in and out at will ... [T]he mind has two main functions; the first one ... is to intellectualize or rationalize ... the second one is to perceive reality directly with no interference of any kind of thinking, to intuitively catch the reality - not what we think it is - see what it is. (1993)

The Actor, then, is seen as a hierophant or shaman with the technologies to make a bridge from the 'shattered universe' into the unified ground of being.

In Núñez's work we find an emphasis on the triad of sacrality, being and cosmos, which our sources have identified as the fundamental religious imperatives. How then should we understand an activity such as that of the TRW, which divorces anthropocosmic technologies from their religious contexts and espouses a secular sacrality?

In *Das Heilige* ('The Holy' or 'The Sacred') Rudolf Otto introduced the term 'numinous' to describe the experience of the 'sacred'. His *mysterium tremendum* is the experience of awe, of overpowering presence and energy, and of enrapturing fascination in the face of that which is 'other'. For Otto, the experience of the sacred is one to which humans are predisposed. That predisposition can find expression in - or be projected onto - the natural world or an imputed supernatural world. Thus, for Otto, the numinous experience is a natural capacity, prior to and not by necessity associated with religion *per se*. The experience of the sacred - or the impulse to imbue aspects or artefacts of experience with sacred connotations - is not exclusive to the domain of religion (Eliade 1957).

One of the primary functions of religion, however, according to Durkheim, is to create a division between the sacred and the profane. His sense of

the sacred is not synonymous with the divine since there are primary forms of worship of the sacred that do not involve divinities, and it was one such (indigenous Australians) on which Durkheim based his explorations. While ultimately Durkheim sees the totem as symbolizing a sacred reality, which is itself a projection of a social reality, it is pertinent to note that the sacred has a fundamentally non-religious connotation for him (1995 [1912, 1915]). William James notes that

> [t]here are systems of thought which the world usually calls religious, and yet which do not positively assume a God... the Buddhistic system is atheistic. Modern transcendental idealism, Emersonianism ... Not a diety *in concreto*, not a superhuman person, but the immanent divinity in things, the essentially spiritual structure of the universe, is the object of the transcendentalist cult.
> (1960 [1902]: 50)

'Secular sacredness' might sound, at first hearing, like an oxymoron. Yet the opposite of the sacred is not the secular but the profane. The concept of the sacred forms a polarity with the concept of the profane, each defining and delineating the other. It is worth noting, as Eliade does, that the poles do not equate to a positive and negative; each may contain both - the realm of the sacred consists of evil as much as beneficent powers. The secular pertains to that which is not the religious life or order, thus to permit a secular sacrality, we must only allow that the sacred may be found beyond the bounds of that which we determine to be the domain of religion. As our definitions above suggest, we might well expect to experience the absolute, the wholly other, the interconnected universe without an organizing principle in the form of religious structure or belief.

Before we leave our discussion of the parameters of the 'sacred', it is worth noting that for Durkheim the sacred also referred to a context greater than the individual and the mundane - the collective life of the community (1995 [1912, 1915]). This will be discussed further below, where we shall see the biological benefits accruing to a practice of service to the community, and shall relate this to the group-basis of Núñez's practice and to the development through that of communitas.

EXPERIENCES IN CONSCIOUSNESS

We have then, a definition of the sacred that allows for secular contexts and emphasizes modes of ontological being, phenomenological states of perceiving and of cognizing the cosmos in which we exist. Experience of the sacred, then, may be seen as an experience in consciousness. Indeed, there is an emergent field of study of precisely this - the consciousness of religious experience - a field that:

1. gives us a framework for analysis (below)
2. establishes the pre-cultural aspects of spiritual practice and experience, which support Núñez's hypothesis
3. provides an insight into the kinds of methodological research that pertain to analysis of religious experience (and, indeed, to the construction of such experiences, and therefore provides a support for Núñez's methodology).

For Andresen and Forman, editors of 'Cognitive Maps and Spiritual Models' (a themed edition of the *Journal of Consciousness Studies*), religious and spiritual experiences may be seen 'not solely as cultural phenomena but as phenomena that can be related to human physiology, and a kind of pan-human technology of human spiritual development' (2000: 7). Consciousness studies allow for neurological and phenomenological perspectives on the modes of being and states of experience that we associate with the sacred. It is these perspectives that better allow us to understand the complex interplay of culture and biology in the emergence of religious states. '[C] onsciousness' stands as the mediating term between the qualia, or felt experience, of the subjective, and the 'hard' reality we refer to as 'the external world' (8).

Andresen and Forman point out that causal vectors move in both directions between consciousness and culture. Cultural constructs including language influence our ability to have specific experiences, but individual subjective experiences also shape culture: 'Culture and consciousness interact with, and reflectively influence, one another, and so do biology and consciousness' (9). This essay will assume both the viability (indeed necessity) of a methodology that incorporates subjective experience and the primacy of measurable physiological and psychological factors within even the most mystical of religious states.

For Perennial psychologists, mystical experiences of sacred reality across cultures 'share certain common underlying experiential cores'. Andresen and Forman specifically identify those instances as 'non-dualistic', as 'largely, or perhaps even entirely, unconstructed by cultural language and background' and as 'unformed' (8, 12).

The emerging picture of religious experience is of 'a particular kind of consciousness' which possibly 'reflects pan-human correlations at a deeper level than conceptuality - electrical activity in the frontal and temporal lobes of the brain, the stimulation of hormone flows and the ceasing of random thought-generation all may be seen as cross-cultural technologies of spiritual experience' (13). At a physiological level, these are 'unformed' states, but how these states are experienced, conceptualized and articulated will largely be shaped by culture and language.

We might, then, understand mystical and shamanic ritual practices as 'technologies' of mental and energetic means by which to generate 'religious' or cosmological consciousness. These are Eliade's 'technologies of the sacred', which Stanley Krippner identifies as 'a group of techniques by which practitioners deliberately alter or heighten their conscious awareness' (Krippner 2000: 98).

Núñez's cross-cultural researches have explored precisely this terrain, as a result of which the 'tools' of the TRW are a set of psycho-physical techniques that can be combined in different ways, and with different mythological imageries, to create the interactive 'dynamics'. The 'tools' include: slow walking, whirling, 'contemplative running', corporeal alphabets and energetic positions. The dynamics are constructed around four fundamental principles, also derived from the research sources: continuous movement; continuous mental focus on one's experience; changing, specific rhythms; alternation between tension and relaxation. The intricate interplay of mental and physical strategies in the dynamics is such that we can describe them as techniques of 'meditation in movement' (Middleton 2001: 46-56).

FROM THE PROFANE TO THE SACRED

Arthur Deikman's analysis of mystical technique delineates three useful polarities within which we can locate Núñez's practice:

> Instrumental consciousness - Receptive consciousness
> Object consciousness - Unboundaried sense of self
> Survival self - Spiritual Self

These in turn highlight three important factors for consideration: intention, attention and perceptual deautomatization (2000: 75-99). Deikman tells us that

> there are different modes of consciousness to serve our basic intentions - they are functional. To act on the world requires a sharp discrimination between self and others and between self and objects... an acute sense of linear time is needed ... In contrast, to take in, to receive from the world calls for a different mode of consciousness, one in which boundaries are more diffuse, the Now is dominant, and thought gives way to sensation. (79-80)

Deikman points out that normal human cognitive maturation involves the development of capacities for object-recognition, boundary-perception and conceptualization. These capacities and their related skills (with regard to space, time, causality and self) become automatic in normal childhood development. Meditative and mystical practice involves, in Deikman's analysis, a deautomatization:

> The meditation activity that my subjects performed was the reverse of the developmental process: the percept... was invested with attention while thought was inhibited. As a consequence, sensuousness, merging of boundaries and sensory modalities became prominent. A *deautomatization* had occurred. (77)

Deikman's meditators were intentionally shifting their consciousness from a daily mode, which has evolved for instrumental and survival purposes, to a 'spiritual' mode of receptivity. The mental apparatus of instrumental consciousness itself creates a 'barrier to experiencing the connectedness of reality'. In contrast, a shift to a receptive mode of being enables both perception and experience to shift also. Deikman identifies that this is fundamentally 'a shift in intention away from controlling and acquiring and toward acceptance and observation. The emphasis is on taking in instead of acting upon' (78–80).

The elements identified by Deikman are common to many meditational practices: focused attention upon a percept, the intention to accept and observe [which in turn implies a cessation of ego-activity], the quietening of conceptualization, the dilation of sensory and perceptual activity. In Núñez's dynamics, attention is focused in the moment-by-moment somatic experience through intentionality, breathing technique or use of mantra. Receptive consciousness is engaged through the necessity to remain within long-durational activities, abandoning end-gaining strategies and time-consciousness. Conceptual activity is subdued, partly through intention, and partly through the psycho-physically strenuous tools of running, energetic position etc. Energies are dilated through physiological effects (such as adrenalin and endocrine release), and this in turn intensifies the somatic nature of the experience.

Deikman has already identified the perceptual deautomatization involved in the shift from one mode of consciousness to another. In Núñez's dynamics we also find a complementary deprogramming on the physical plane. Many of the dynamics involve physically counter-intuitive actions, such as running backwards with eyes closed, or some of the complex turns in *Citlalmina*. This, too, helps the participant to remain in a state of alertness and presence, vigilant to the necessity to maintain a non-habitual attentional energy.

• Nicolás Núñez with members of the TRW dancing *Citlalmina*. *Photographer: Steve Forrest.*

These, then, are psycho-physical structures with the potential to facilitate a meditator's refusal of ordinary consciousness, with its inherent limitations and filtering constructs, in favour of what Deikman calls 'a mode of consciousness responsive to [reality's] connected aspects' (89). Or, rather, a mode of consciousness that does not, by nature of its developmental apparatus, prevent the experience of objectless, unboundaried connection.

SECULAR SACREDNESS

It seems, then, that we may access what we are calling 'religious' states of consciousness, in which to experience cosmos, simply by utilising technologies of mind and body. We must be mindful, however, that those states will be experienced and conceptualized in ways that depend upon the cultural and language backgrounds of the subjects. Karoliina

Sandstrom, who has for two years been participating in the experiences of *Citlalmina*, which I run at the University of Huddersfield, writes:

> So for me there exists a quest for 'being' in the universe, which I will approach from a particular individual and societal stance. Even though each individual's understanding of 'being' differs, the practice of, for example, *Citlalmina*, facilitates a group participation in which there is the possibility of such a quest taking place. And for me this quest for 'being' (or warriorship) can in some degree transcend cultural and societal contexts to allow for possibly very different yet still related spiritual experiences. (Middleton 2008)

In 2006, Núñez led a group from the TRW in a 'high-risk theatre' training exercise comprised of ritually walking the *camino*, or pilgrimage, to Santiago de Compostela in Northern Spain.[4] Lee Rickwood, an Australian actor who joined the company for that experience and went on to work with them in the subsequent theatre production, *Cacería de Estrellas* (2007), describes the ways in which the group were facilitated in accessing and understanding the experience as a secular pursuit:

> The approach that Nicolás had devised for the group was held together by the image of Quetzalcoatl, the feathered serpent ... The major historical 'story' of the Camino is that it is a Christian pilgrimage ... Yet walking in the serpent allowed us to have an experience of the Camino outside this dominant narrative, creating our own, shared, secular sacred experience. We were a group joined not by a religious order or set of beliefs, but by our imaginations ... and in walking it becomes a psycho-physical sacred experience.
> (Middleton 2008)

In Rickwood's analysis, both personal imaginative response to archetype and the psycho-physical effects of ritual walking facilitated a numinous experience, which she defined as entirely secular, divorced from the 'dominant narrative' adhering to it. Eilon Morris, who has experience of the dynamics both with Núñez and in my use of them in England, describes his practical strategy for encountering the forms in a way that acknowledges cultural contexts:

> My personal experience is that through approaching these forms with humility and respect and an understanding of some of the principles that underlie them, I am able to engage in a personal relationship between myself and the experiences revealed through their practice. This engagement occurs without an in-depth knowledge or personal connection to the cultures that they are drawn from. There is the risk, I am aware, that if the forms are performed purely as physical movements and body positions, they may only function as physical exercise and or dance moves. Therefore I feel there is a need for some form of personal connection and commitment to the work. (Middleton 2008)

As with meeting another person, we can have an authentic encounter with a performance form, even across cultural, experiential and linguistic divides, when that encounter is forged in humility, respect and a pre-conceptual level of engagement. Morris's principles, derived from work with Núñez and others and through which he accesses the dynamics, orient him to an approach that reduces the risk of cultural imposition and facilitates receptive consciousness in Deikman's terms. Morris writes:

> - a vigilant approach to working with my attention in order to support my intentions in the work
> - ongoing search for the living potential within the form
> - finding and drawing on a personal inner mantra
> - relaxing my vision, allowing it to move freely throughout the work
> - listening to the sounds and rhythms of the rattles and the steps and allowing these to work through me, rather than attempting to control them. (Middleton 2008)

It is clear that psycho-physical forms are constructed not purely from physical actions but from the interplay of those actions with highly specific attitudinal foundations. Núñez's

4 I joined the TRW halfway through their forty-day pilgrimage to Compostela and, beyond it, to Finisterre on the west coast, a point considered by some to be the end of an ancient pre-Christian pilgrimage into the West.

• **Members of the TRW walking the Camino to Santiago de Compostela as a 'high risk theatre' training.** *Photographer: Arie van Duijn.*

'meditation in motion' requires the same disciplining of mind as can be found in other forms of sitting or walking meditation. Ultimately, indeed, the interior orientation is the most important element. As Deikman explains,

> [a] student can meditate for years focused on breathing sensations but if she is inwardly trying to grasp enlightenment, to possess it, that acquisitive aim will lock her into the same form of consciousness with which she had begun. That is why mystics say that 'the secret protects itself'. No cheating is possible because it is the interior orientation that is critical. (2000: 78)

One might say, too, that appropriation of sacred forms through acts of cultural piracy or vandalism are also protected against in this way. It is only when engaging with an authentic interior attitude that a person can be said to be actually performing the form and, indeed, benefiting from it. Otherwise, as Morris suggests, it is simply 'physical exercise'.

The meditative, or spiritual, path is itself a vehicle for moving towards the attitudes of inner-orientation that are crucial here. Andresen and Forman describe the ways in which such practices centrally consist of an incremental movement away from culturally formed consciousness to the unformed pan-human levels of mind activity mentioned earlier.

> One begins a spiritual practice, of course, utterly enmeshed in the historical world ... As meditation slowly moves one away from sensation and thought, the formative role of background and context slowly slip away. One becomes less aware of one's surroundings, thinks with borders that are less and less defined ... [I]n some branches of Buddhism and Hinduism, practitioners are believed to have become less and less enmeshed in samsara, the cyclical mundane world. (2000: 12)

This, however, brings us to a critical distinction between experiences of altered states of consciousness on the one hand and incremental processes of spiritual development on the other.

STATES AND STRUCTURES

According to Ken Wilber's extensive survey of religious experience in Eastern and Western psychological systems, 'a person at virtually any stage or level of development can have an altered *state* or peak experience - including a spiritual experience' (Wilber 2000: 149). When these are developed into 'permanent traits' to which the individual has 'more-or-less *continuous* and conscious' access, 'these transitory states are converted into permanent structures of consciousness' (150, 154). Access to altered states of consciousness appears to be an innate capacity, but the related structures of consciousness, which would enable these states to become permanent traits, must be developed through a staged evolution (165, 153). Further, Wilber points out that

> the ways in which these altered states will (and can) be *experienced* depends predominantly on the

structures (stages) of consciousness that have developed in the individual. (154)

While I have proposed that Núñez's practices meet the criteria for creating conditions for the arising of spiritual states of consciousness, we must now ask whether the work also provides for an ongoing developmental process through which those temporary states could influence the individual's ongoing experience. If a practice facilitates the arousal of states but does not provide a developmental process, what are the implications for that practice as a spiritual vehicle? As a means of providing access to 'secular sacredness'?

For Helena Guardia, Núñez's long-term collaborator in the TRW, these are also central questions:

> What's the use of dynamics and practices that take you so energetically to the right side of your brain, that raise your energy so powerfully and make it soar, that open new scopes and give a deeper and more joyful meaning to life, if we do not extend this and produce a real change in our daily affairs. These practices put us in contact with our Being, that place where we can feel that it might be true that we are one with the 'ten thousand things' ... If you practice but cannot control your depressive emotions, is it useful? What's the use of ritual if we cannot 'transport' this dimension into our everyday lives, in order to 'transform' them?
> (Middleton 2008)

In what ways, then, does the work of the TRW encourage not only immersion in transient being states but the development of those states into altered structures of consciousness? Robert Forman explains that

> [the] discriminating feature is a deep shift in epistemological structure: the experienced relationship between the self and one's perceptual objects changes profoundly. In many people this new structure becomes permanent. (1998: 186)

Thus, while the wisdom literature of the world records numerous examples of sudden 'Road to Damascus' experiences producing permanent epistemological changes, we might also expect to find long-term repetition of practice, as well as incremental internal processes of engagement. I have previously identified within the TRW's dynamics a series of practical objectives – like a ladder, 'each rung facilitating access to the next. The participant's attainment of each rung is also ... the attainment of a level of psycho-physical development':

> 1. silence the rational mind; control the mind's tendency to engage in discursive and instrumental activity
> 2. be present in the here and now and focus on the somatic sensations of the bodymind
> 3. focus mental and physical energies and work with raising and controlling energy (Middleton 2001: 50)

At each of these levels, the participant enters into relationship with aspects of the self (thoughts, feelings, energies), thereby dissolving habitual identification with that aspect and developing the 'observer self' of meditative

• Nicolás Núñez with the vibratory conch in the TRW production, *Cacería de Estrellas* (Mexico, 2008). *Photographer: Marco Lara*

practice, which enables a profound epistemological shift away from ego-consciousness into the active receptivity that Deikman identified as crucial to the mystical path.

For Núñez, actor training requires an incremental process akin to that of the spiritual path. He writes that

> [t]he training tools of a true actor go hand in hand with the training tools of the spiritual warrior ... [T]he actor has to train his mind, his speech and his body through discipline, intent, concentration and intelligence, exactly as, for example, the Tibetan Shambhala-tradition warrior, or the warrior from the Meso-American Nahuatl tradition ... Both are disciplines of spiritual warriors with the same aim of becoming channels for the sacred.
>
> (Núñez 2000: 21)

Thus it is that the activities of the TRW attract both actors in training and individuals seeking personal and spiritual transformation, both Mexicans making contact with their own cultural heritages and non-Mexicans for whom the work provides structured and accessible experiences at a pre-cultural level.

THEATRE'S SACRED DIMENSION

> Theatre is literature and spectacle, play and catharsis, but it is also an initiation ceremony. Núñez and his colleagues have striven to re-introduce the sacred dimension into theatre.
>
> (Octavio Paz in Núñez 1996: xx)

Núñez sees the history of theatre as a development away from ritual efficacy and towards entertainment (Núñez 2007: 14; see also Schechner 1988: 106-52). He tracks a historical trajectory from early ritual contact with 'pure energy' into the conversion of that energy into deity, and from there into the personified epic culture hero. When the culture hero 'becomes a human being... he performs no more extraordinary feats and now he has emotional conflicts; his rank is theatrical' (Núñez 2000: 11). The history of theatre, for Núñez, is one of descent from the sacred to the profane.

The work of the TRW has been to reinvest contemporary theatre with ritual potentialities and sacred connotations, to move from a foundation in theatrical activity into the realms of the epic and the mythic and from there to the experience of pure energy and cosmos that we associate with archaic rites. In Núñez's analysis, contact with the sacred is the birthright of the theatre, and the actor and hierophant are like twins who share common energetic and imaginal capacities to access other realms of being, other levels of consciousness (2007: 17). In recent history, Western theatre has aligned itself with commerce and the profane; only in some isolated examples (notably that of Grotowski, with whom Núñez trained and worked [Núñez 2000: 51-64]) do we find the Actor using their trained potentials to access cosmos.

And yet, there are ways in which the core characteristics of theatre continue to provide a natural environment for a sacred pursuit. Helena Guardia writes,

> It is possible to enter altered states of consciousness through any form of art, all of them take you beyond time and space, as meditation does, but only in theater - as ritual (for it involves participation) - can you share the precise, living instant of soaring, and the mutual commitment deriving from it. (Middleton 2008)

As a fundamentally communal art form, theatre provides a space in which one's meditative and energetic training may be supported by a congregation-like body of mutual initiates. This is important in two ways; one, encounter with the 'other' facilitates the pursuit of 'being' mentioned above, and two, the group experience of communitas supports individual effort and equates to what Deikman saw as a biologically adaptive characteristic of spirituality: 'service'.

On the role of the other in ones individual journey, Karoliina Sandstrom tells us

> [t]here is a strong sense of shared experience, which, for me, not only relates strongly to the possibilities of developing one's engagement in

> performative work, but also suggests another possibility of engagement in the universe ... There are times in the work when I feel that the dynamic and the group are helping me to meet myself and the world ... I see a strong link between the work and ... meeting [another] person ... and being present in ones performative practice. (Middleton 2008)

Many of the TRW's dynamics place the individual in profound contact with another. In the midst of techniques that shift our identification away from ego-self, meeting another person can represent a rare opportunity for undefended presence - being - which we see reflected in the partner's eyes (Middleton 2001: 58). Often, the contact with the other involves facilitating their experience, as many of the dynamics include tools, such as running backwards with eyes closed, that require the presence of a vigilant partner. The TRW's aim may be to provide rituals of personal transformation, yet the work can never be individualistic or self-centred. One operates within the group as within a theatrical ensemble, fitting ones energies and actions to the group needs and relying on that group, in the long-durational actions, for energetic support. This creates a strong sense of communitas, even among large groups of strangers meeting in the work for the first time.

In dynamics such as *Tloque Nahuaque*, participants move in repeating rhythmic patterns of coming together and moving apart, in such a way as ultimately to blur one's sense of boundaries and separateness from the other. Lee Rickwood describes her experience:

> I moved from a sense of separateness to a sense of connectedness to a sense, midway into the dynamic, of an ongoing interconnectedness, whether in the cluster or moving separately. For me, this experiential interconnectedness was an experience of communitas and sacred space ... energizing, expansive and empowering. (Middleton 2008)

As Deikman has noted, mystical practices 'help a person "forget the self", to diminish the extent to which survival needs dominate consciousness' (2000: 83). We have already considered the role of meditation in doing this; Deikman also offers an analysis of the practices of 'renunciation and service', each of which dissolves self-centredness and encourages a receptive mode of being and an experience of interconnectedness. The aim of the TRW is to create experiences in which participants can align themselves with the cosmos, and this is achieved through the cultivation of a strong internal discipline, necessary for 'serving the task' (what Deikman calls 'true service' [85]). It is for good reason that Antonio Prieto Stambaugh draws attention to the 'collective nature of much of the [TRW's] activity' (2002: 7).

With his collaborators in the TRW, Núñez has carried out a phenomenological research across cultures in order to establish the parameters and principles of an anthropocosmic theatre. The result is a paradigm forged upon psycho-physical practices, which are supported not by religious ideas or doctrines, nor by conceptual ideologies, but rather by technologies of attention and intention, modes of relation to self and other.

REFERENCES

Andresen, Jensine (2000) 'Methodological Pluralism in the Study of Religion', *Journal of Consciousness Studies* 7.11-12: 17-73.

Andresen, Jensine and Forman, Robert (2000) 'Methodological Pluralism in the Study of Religion', *Journal of Consciousness Studies* 7.11-12: 7-14.

Barba, Eugenio (1991) *A Dictionary of Theatre Anthropology: The Secret Art of the Performer*, London: Routledge.

Deikman, Arthur (2000) 'Methodological Pluralism in the Study of Religion', *Journal of Consciousness Studies* 7.11-12: 75-91.

Durkheim, Emile (1995 [1912, 1915]) *The Elementary Forms of Religious Life*, New York and London: Free Press.

Eliade, Mircea (1957) *The Sacred and the Profane*, San Diego, New York and London: Harcourt.

Eliade, Mircea (1989 [1964]) *Shamanism: Archaic Techniques of Ecstasy*, London: Arkana.

Forman, Robert (1998) 'What Does Mysticism Have to Teach Us About Consciousness?' *Journal of Consciousness Studies* 5.2: 185–201.

Huxley, Aldous (1944) *The Perennial Philosophy*, New York and London: Harper.

James, William (1960 [1902]) *The Varieties of Religious Experience*, London and Glasgow: Fontana.

Krippner, Stanley (2000) 'Methodological Pluralism in the Study of Religion', *Journal of Consciousness Studies* 7.11-12: 93–118.

Light, Timothy (2004) 'Orthosyncretism: An Account of Melding in Religion' in Anita Maria Leopold and Jeppe Sinding Jensen (eds) *Syncretism in Religion*, London: Equinoz, pp. 325–47.

Middleton, Deborah (2001) 'At Play in the Cosmos: The Theatre and Ritual of Nicolás Núñez', *TDR* 45.4: 42–63.

Middleton, Deborah (2008) Unpublished interviews and personal correspondence.

Núñez, Nicolás (1993) Conference address at 'Performance, Ritual and Shamanism', Centre for Performance Research, Cardiff, Wales.

Núñez, Nicolás (1996) *Anthropocosmic Theatre: Rite in the Dynamics of Theatre*, Amsterdam: Harwood.

Núñez, Nicolás (1997) Personal correspondence with the author.

Núñez, Nicolás (2000) *The Flight of Quetzalcoatl*, Unpublished pamphlet.

Núñez, Nicolás (2007) *Teatro de Alto Riesgo*. Mexico City: SIGAR.

Otto, Rudolf (1923 [1917]) The Idea of the Holy, trans. John W. Harvey, London: Oxford University Press.

Prieto Stambaugh, Antonio (2002) 'To the Editor', TDR 46.3: 7–8.

Schechner, Richard (1988) *Performance Theory*, New York: Routledge.

Stewart, Charles and Shaw, Rosalind (1994) *Syncretism/Anti-Syncretism*, Oxon: Routledge.

Wilber, Ken (2000) 'Methodological Pluralism in the Study of Religion', *Journal of Consciousness Studies* 7.11-12: 145–76.

Liturgy and Mass Spectacle

The case of Catholic mass theatre in Flanders during the interwar period

THOMAS CROMBEZ

It is a well-known fact from modern theatre history that, when Leon Trotski attended *Earth Rampant* at the Meyerhold Theatre in 1923, he suddenly appeared on the stage to deliver a speech on the fifth anniversary of the Red Army. Since the play by Sergei Tretyakov was explicitly concerned with the events of the Civil War, his address fit in perfectly. 'After thunderous applause the action on stage continued as if without interruption and Trotsky again returned to his seat' (Yuri Annenkov quoted in Leach 1993: 19).

Trotsky's interruption marks a key event in the history of the theatrical avant-garde. According to Peter Bürger's influential *Theorie der Avantgarde* (1974) - and numerous similar texts - its main project was to reunite art and the everyday praxis of life. In Bürger's concise words, the utopian aim of avant-garde artists could be phrased as 'the attempt to organize a new life praxis from a basis in art' (Bürger 1984: 49). Consequently, the fame of Trotsky's intervention is even outshone by Soviet mass spectacles such as *The Storming of the Winter Palace* (Petrograd, 1920). The enormous production commemorated a decisive event from the Revolution. Here the aesthetic happening coincided almost seamlessly with the social. Many of the eight thousand 'actors' and the hundred thousand spectators had been involved, a few years before, in the actual historical events. Moreover, the Civil War was still raging in the vicinity of Petrograd at the very time of the show. Present-day evaluations, by Susan Buck-Morss and Slavoj Žižek, continue to stress the event's avant-garde character, for example by quoting a contemporary report: 'The future historian will record how, throughout one of the bloodiest and most brutal revolutions, all of Russia was acting' (Žižek 2003; Buck-Morss 2001: 144).

Revolutionary mass theatre has become a well-known and even paradigmatic example of avant-garde art. In this essay, I would like to explore other types of mass theatre or, to use a more general expression, 'socio-theatrical events'. My case-study will concern the mass spectacles and choral plays organized by Catholic reformers during the interbellum period in the region of Flanders (Belgium). Although we do not commonly associate such events with the revolutionary fervour of the avant-garde, they do fit Bürger's description of the avant-garde project surprisingly well.

My guiding question will be that of *liturgy*. I would like to investigate the properly 'religious' dimension of these events. Were they modelled on traditional Catholic liturgy (Mass, the sacraments), or did they rather fit in with a modernist aesthetic? As these events were organized by the Catholic Church but directed by modern theatre artists who were keenly aware of artistic currents such as Constructivism and Expressionism, I believe that the Flemish mass spectacles of the 1930s form a highly interesting case of 'arrière-garde' (Marx 2004).

1. MASS THEATRE IN EUROPE

Mass theatre was a popular expression of 'community art' in most European countries

Performance Research 13(3), pp.55-63 © Taylor & Francis Ltd 2008
DOI: 10.1080/13528160902819323

during the interbellum period. From Max Reinhardt's lavish open-air spectacles (such as *Everyman* in Salzburg, 1920) to Socialist workers' *Laienspiel* (lay theatre), theatre visionaries focused on ever larger groups for entertainment as well as political agitation. Among Western European countries, Belgium was one of the last to follow the new trend of socio-theatrical events. It was imported via the Socialist movement from Germany and Russia, passing through the Netherlands, where workers' choirs (both singing and reciting choirs had quickly risen in favour.

In the case of Belgium, and Flanders in particular, it is especially interesting to observe that the mass theatre phenomenon displayed an ideological heterogeneity not seen elsewhere. Catholicism soon appropriated the Socialist idea of lay theatre and choral training. Their spokesmen defended a modernization of Catholicism that was at the same time traditionalist and radically contemporary. How did they come to conceive the renaissance of the ancient liturgical tradition in the form of mass spectacle?

• **Jozef Boon, Reciting choir *Noodsignalen* (Signals of Distress) in Esschen (1934)**

I will first study the emergence of liturgy in general in modern drama theories of the early twentieth century. This will serve to highlight the peculiarities of Catholic Flemish theories by directors and critics such as Jozef Boon, Lode Geysen, Aloïs de Maeyer and Herman de Vleeschauwer. The concluding movement of this essay will be devoted to verifying whether the liturgical component was not simply an important theoretical presence. Was liturgy equally strongly articulated in the actual stagings of Catholic mass theatre? Was it possible to successfully merge Catholic liturgy with a modernist theatre aesthetic?

2. LITURGY AND MODERN DRAMA

Although the Soviet director Meyerhold was an ardent promoter of revolutionary mass spectacle, he sharply realized that the conservative forces in society had long since employed the means of theatrical street actions. Far from blind to the new impulses which emanated from the proselytizing 'Catholic Action' movement (stimulated by Pope Pius XI), he wrote in 1929:

> The Vatican has become a laboratory for research into the art of production. Of all theatre-directors, the Pope is the most inventive, the most ingenious. Even now, as I sit checking this summary of my lecture, the newspaper contains a report of widespread anti-Semitic disturbances which have turned into an organized Jewish pogrom. But take note: 'On the eve of the disturbances in Lvov there were organized Catholic processions' ... Now the Catholic Church has offered its hand to Fascism. With their religious processions and their Fascist rallies, these two organizations are unrivalled in their restoration of the traditional devices of the street theatre. (Meyerhold 1969: 260)

It was not only political parties and religious organizations that flirted with the social and dramatic potential of liturgy, but also many diverse contributors to modern drama. The most caustic among them was T. S. Eliot, who, in 'A Dialogue on Dramatic Poetry' (1928), wrote: 'The only dramatic satisfaction that I find now is in a High Mass well performed' (Eliot 1972: 47). His experiments in liturgical drama would first lead to the pageant play *The Rock* (1934) and subsequently to *Murder in the Cathedral* (1935). In the same vein, the French dramatist Paul Claudel came to reproach his modernist idol Arthur Rimbaud for attempting to approach Eternity through the means of the senses, i.e., secular poetry. Being merely a 'poet without the power of the priest', Rimbaud failed to realize that only the unleavened bread of the Eucharist could be a material pointer to the eternal realm (Claudel 1957 [1919]: 500-1). Claudel's artistic conclusion was to introduce liturgical elements in the texture of his plays. He composed the final redemptive scene of *Partage de midi* (1906), for instance, by means of a litanical and confessional language (aptly entitled 'Mesa's Canticle').

In *Le Masque et l'encensoir* (1921), Gaston Baty developed a similar intuition into a historical essay on the origins of theatre. Religious ceremony, and most importantly medieval Catholic liturgy, was the true dramatic phenomenon, because it had integrated all dramatic components - the spoken text and the non-verbal elements of spectacle - into a harmonious whole. Classicist drama had corrupted the dramatic harmony by privileging the spiritual and individualist dimension, i.e., the text. Baty undertook to rewrite Western theatre history as a fluctuation between these two poles of the theatrical event: the spiritual dimension of drama, which had dominated literary theatre since the Renaissance, and the material dimension of spectacle, which had resurfaced in such genres as commedia dell'arte, classical ballet, melodrama and pantomime. Only the distinctly Catholic and medieval creation of Mass - the ineffable divine mystery attired in the material splendour of liturgy - had succeeded in harmonizing both poles.

For Baty, though, a contemporary Catholic theatre did not mean an art-form explicitly intended to edify or apostolize. 'The dramatic system of the Middle Ages does not at all require a religious topic. The same living and elastic form may serve a romantic intrigue or a historical action' (Baty 1949: 87). Here we touch on the true problem that liturgy posed to modernist theatre practitioners. Convinced that theatre should break out of the autonomy of art and approach the efficacy of ritual, it was unclear how this efficacy was to be understood or realized. Eliot and Claudel attempted each in their distinct way to integrate a certain liturgical style in drama. For Baty, Catholic liturgy stood for a superb harmony between word and image. Others still modernized and restaged the medieval liturgical drama itself, such as Gustave Cohen and Henri Ghéon in France, or Herman Teirlinck and Herman van Overbeke in Flanders. But however great their admiration for its sacred efficacy, few modernists went so far as to strive for a genre truly in-between religious practice and modern drama. They tried to imbue the existing Western theatre with the powers of liturgy but were reluctant to demand the creation of a new dramatic genre that would truly reinvent liturgy for the modern age. Others, however, unhesitatingly stated that such a form could be made. Its creators should be the ideologically organized masses of the twentieth century.

3. MODERNIST CATHOLIC THEATRE IN FLANDERS

Before I contrast the views of Eliot, Claudel and the others with those of the Flemish mass theatre practitioners, it is necessary to frame their activities within the context of interwar Catholicism. The origins of Catholic reciting choirs, which quickly grew into the large-scale spectacles of the 1930s, may be found in two distinct developments of the early twentieth century: Catholic Action and the rise of modernist Catholic theatre.

From the other side of the political spectrum, Socialism had succeeded in organizing the working class by means of trade unions and youth movements. These comprised an important cultural component. Prominent artists such as Bertolt Brecht helped to compose an extensive repertoire of militant songs, *Lehrstücke*, and poems and choral dramas

• Lode Geysen, KAJ Mass spectacle in the Heysel Stadium (Brussels, 1935)

intended for the workers' choirs (amongst which the *Arbeitersprechchöre* or reciting choirs). In Germany and Flanders, young professionals from the budding modern dance scene made use of the Laban system to train Socialist lay movement choirs (notably Laban's disciples Martin Gleisner, Albrecht Knust and Lea Daan).

The Church had responded to the growing popularity of Socialist organizations by stimulating the creation of similar Catholic groups. Especially the 1922 encyclical letter *Ubi Arcano Dei Consilio* consolidated the concept of a society-wide Catholic Action that would 'restore all things in Christ' (following the motto of Pius X) through engaged lay groups. Youth movements, assisted by chaplains, were widely deployed to counter the popularity of Socialism among young workers. Catholic Action found one of its most outspoken and active promoters in the person of Belgian priest Jozef Cardijn. From his 'unionized youth' movement *Jeunesse Syndicale*, founded in 1919, quickly grew the successful Young Christian Workers. Their members were labelled 'Jocistes' in French-speaking Belgium (JOC - *Jeunesse Ouvrière Chrétienne*) and 'Kajotters' in Flanders (KAJ - *Kristene Arbeiders Jeugd*). As my main subject in this essay will be the developments in the Flemish movement, I will subsequently use the acronym 'KAJ'.

The KAJ's foremost task was to incite Catholic unionization of young factory workers. To that end, it provided a Catholic version of Scouting-type activities, but additionally demanded that its members closely monitor the local work force. It was an evident evolution to copy the Socialist youth movements' cultural projects, too. The programme of Catholic Action had already implied a strong revaluation of such theatrical expressions of popular piety as processions and passion plays.

> We refer to the various organizations of young people which have helped to develop such ardent and true love for the Holy Eucharist and such tender devotion for the Blessed Virgin, virtues which have made certain their faith, their purity, and their union one with another; to the solemn celebrations in honor of the Blessed Sacrament, at which the Divine Prince of Peace is honored by truly royal triumphal processions. (Pius XI 1922: 53)

These traditional forms of popular piety would not only be intensified but also supplemented by new and modernist devotional events. Reciting choirs and movement choirs, composed of amateurs, were trained by theatre and dance professionals in order to participate in the mass happenings that celebrated the Catholic renouveau of the 1930s. The largest and most notable events include the mass play at the Heysel Stadium in Brussels by the KAJ, directed by Lode Geysen in 1935, the Rerum Novarum play *Bevrijding* (*Liberation*, Antwerp, 1936) and Jozef

Boon's *Sanguis Christi: The Play of the Holy Blood* (Bruges, 1938).

If Catholic Action may account for the religious and socio-political origins of these events, it says little to nothing about their artistic sources of inspiration. During the first decades of the twentieth century, a torrent of innovations in stage design, dramaturgy and acting had been introduced by the predominantly left-wing theatrical avant-garde. Constructivist, Expressionist and Futurist artists had demanded that art conquer the public space and let itself be conquered by it. To that end they employed innovative drama texts that were composed by montage or featured actual events. Popular theatre forms inspired new acting styles, such as Meyerhold's biomechanics or the Futurists' provocative delivery. Large groups of performers were deployed in novel ways, e.g., the chorus equipped with megaphones for Ilya Selvinsky's *Commander of the Second Army* (directed by Meyerhold in 1929). The stage itself was invaded by banners with revolutionary slogans, giant constructions, everyday objects and projection screens.

In Flanders, the Vlaamsche Volkstooneel (Flemish Popular Theatre) was one of the few European ensembles that strove to integrate modernist techniques in a traditionalist, Catholic framework. Constructivist scenography and stylized acting helped to stage plays based on figures of Flemish folklore, such as *Tijl* (Anton Van de Velde, 1925). From the early 1920s, its directors also started to experiment with bigger groups of performers, beyond the customary crowd scenes. The search for an original way to stage the choral songs in tragedies by Sophocles and Vondel eventually led Renaat Verheyen and Lode Geysen to autonomous choral plays. Geysen, along with teacher priests such as Jozef Boon and Gery Helderenberg, soon started to form small reciting choirs with members of the KAJ and other Catholic youth movements. Next, a series of Catholic Action workshops helped to diffuse the newly developed method among Catholic teachers. The end result was a surge of lay-theatrical enthusiasm, animated by techniques from modernist theatre praxis, that produced the gargantuan events from the late 1930s.

4. FLEMISH THEORIES OF LITURGICAL DRAMA

The practitioners of Catholic lay theatre obviously had good reasons to develop a coherent theory of choral drama. The phenomenon was brand new, and it ought to be introduced quickly and convincingly among Catholic groups and theatre professionals alike, if the movement was to gather any momentum. At the same time when the first choral play-texts appeared in print, a series of theoretical manifestos was published by the same small but prolific body of writers (director Lode Geysen, writer-director Jozef Boon, and critics Aloïs de Maeyer and Herman de Vleeschauwer). As was to be expected, liturgy fulfilled an important role in their reflections.

It is conspicuous that the Flemish theorists did propose a much more radical view of liturgical drama than the other modernists treated above, such as Eliot and Claudel. If we apply Bürger's definition of the avant-garde strictly, Catholic lay theatre ought to be called avant-gardist. It literally strove to found a new, impassioned praxis of (Christian) life from a basis in 'art', i.e., choral drama and mass spectacle. Boon surmised that the recent evolution of modern drama, and he was indeed thinking of choral drama, had led it to fuse with 'the chief drama - life - Catholic life with its Catholic tragedy' (Boon 1937: 57). Geysen strongly echoed the Futurist definition of theatre when he asserted that, during choral plays, the action takes place in the auditorium.

> The spectator is a crucial element of choral drama, who has to be incessantly taken aim at ... It is a confession of faith that petitions the audience. All doubt, all hesitation has to be overcome.
>
> (Geysen 1934: 41)

Similarly, when Geysen concluded that the lay player is no actor but only plays himself, this is a verbatim prefiguration of Žižek's typically avant-gardist appraisal of *The Storming of the Winter Palace* as an aestheticization in which the people plays itself. '[O]ne should perceive, in this minimal, purely formal, difference of the people from itself, the unique case of "real life" differentiated from art by nothing more than an invisible, formal gap' (Žižek 2003). The Flemish nationalist critic De Vleeschauwer saw choral drama as a symptom of the contemporary era's inner need for liturgy. He insisted that 'from the art of choral drama must grow the liturgy of labour, freedom and humanity' (De Vleeschauwer 1931b: 172). 'Socialism' - although he was probably thinking of National Socialism - should be instrumental in this process. It should *consciously* cultivate the workers' need for 'irrational mythical and mystical forces'. De Vleeschauwer is clearly echoing such doctrines as Nietzsche's philosophy of 'vital lies', or the French sociologist Georges Sorel's on the re-emergence of myth as a guiding force for modern social movements. In Germany, a similar train of thought would lead to the National-Socialist *Thingspiel*. This, indeed, is one of the possible outcomes of socio-theatricality, and also of the avant-garde: the use of mass theatrical art to refashion the structure of society.

Next to its avant-gardist radicality, the *history* of choral drama was a second important component of the Flemish theories. Since the invention of modern choral drama was usually claimed by its Socialist and mostly German promoters, this was a crucial point indeed. Catholic theorists needed to prove, first, that choral drama had 'always been there' in the history of Western theatre and, secondly, just as much that its modern revival was due to their independent efforts. For the first appearance of choral drama, all pointed to Greek tragedy and its reworkings in the early modern period. But they did not really stress this aspect of the genealogical chain, since the choral songs from ancient tragedy obviously had little in common with their decidedly modernist efforts. Boon, Geysen and De Maeyer would instead underline that the versicles and responses from Mass were the true ancestors of modern choral drama, which was identically structured. The main German theorist of choral drama, Friedrich Roedemeyer, had himself acknowledged this in his book *Vom Wesen des Sprechchors* (1926).

A third remarkable point concerns the question of spontaneity. Although Boon and Geysen, both prolific producers of choral dramas, repeatedly complained that there were not enough suitable play-texts available, they also vividly asserted that a true choral drama could only originate from the group's solidarity and its sense of community. 'Choral plays... can only be staged when the text contains the thinking and willing of the performers' (Geysen quoted in De Maeyer 1933: 115). This is especially striking when the subject of gesture is brought under discussion. Boon and De Maeyer warned that the accompanying gestures, instead of being truly animated, could easily turn into gymnastics when all performers were mechanically instructed to carry out the same movements. Producers therefore had to be attentive for gestures that arose spontaneously out of the group itself, in response to its sense of community or to the choral play that was brought to their attention.

The importance of spontaneity, taken together with the first point - that of the avant-gardist construction of a new praxis of life and hence new communities - indicates a central problem of socio-theatrical events. How orchestrated or spontaneous was the collective life of the groups that actually constituted the ubiquitous 'community' that was worshipped through the interbellum period? A final quotation from Vleeschauwer acutely shows how this tension was alive just under the surface of the discourse of choral drama.

> Modern choral drama is the art of the masses. It is the art that humanity will use to combat that of yesterday and of today. It is an orchestra of human voices, graded and differentiated as organ pipes. It is a living organ, the tones of which are not united by a mechanism and the will of a single human, but by the collective will of all performers.
>
> (De Vleeschauwer 1931a: 91)

5. THE PRAXIS OF LITURGICAL DRAMA

If the spontaneity of choral drama was inherently problematic, the question arises whether the promoters of lay theatre rightfully professed that it was closely related to liturgy. Was 'liturgical drama' a convincing claim? To answer that question, I would like to take a look at a selection of events from the late 1930s.

In his book *Spreekkoor en Massatooneel* (1937), Jozef Boon reported on an interesting and untitled choral drama he directed in the town of Diest with a group of 100 students, on the occasion of the feast of Saint Jan Berchmans, a seventeenth-century Jesuit priest who was the students' patron saint. The action took place on the local marketplace, and Boon had intentionally placed his chorus on a raised podium so that it could 'dominate the market', i.e., the audience of 2000 students that would respond to the reciting choir (Boon 1937: 79). The happening was adequately framed against the imposing backdrop of Diest's Gothic church. The relics of Saint Jan Berchmans were displayed in front of the church portal, and the carillon played specially composed music by Arthur Meulemans, a frequent collaborator of Boon's mass plays.

What is particularly striking about the Diest mass spectacle, is that Boon strove to integrate his choral drama with the official church festivities devoted to Saint Jan Berchmans. The bishop and the clerics who participated were explicitly designated as 'being part of the choral drama' (Boon 1937: 79). Indeed, the event ended with a huge open-air celebration of Mass. If one is to believe Boon himself, the event truly 'conquered' its attendance.

> Immediately the masses were overwhelmed: they were dominated by the fact that carillon, music, church, city hall, flags, podium, and flags and crowd were united into a single body. No small role was played by the wide-circling music that made everything into one wave ... [I]t had grown into a mass happening, especially because it was a part of reality, namely, because of the integration of the Holy Relics and because, lastly, while the choir was still plastically deployed and the musical finale burst open, the priest came with the Blessed Sacrament. All of that was really part of the choral drama, including the blessing by the bishop with the Blessed Sacrament being administered amidst the kneeling choral reciters. (Boon 1937: 79)

One year later, Boon directed what was probably the largest Catholic mass play of the 1930s. *Credo!* ('I believe') was staged in the Heyssel Stadium (Brussels) at the occasion of the Sixth Catholic Congress of Mechelen in 1936. The spectacle numbered hundreds of participants grouped in singing, reciting and movement choirs. A gargantuan audience of 150,000 filled the stadium, usually reserved for sporting events.

• Jozef Boon, Mass spectacle in Diest on the occasion of the feast of Saint Jan Berchmans (1935)

• **Jozef Boon and Arthur Meulemans, *Credo!* (1936), mass performance in the Heysel Stadium (Brussels) to celebrate the VIth Catholic Congress of Mechelen**

Credo ! te Brussel. De Babylon-menschen bij het verschijnen van den engel

• **The traditional penitential procession of Veurne, integrated in the mass performance "Credo!"**

In Boon's eyes, the huge crowd saying the Confiteor prayer, at the beginning of confession, grew into 'the most beautiful and ardent drama ... that we could ever dream of' (1937: 86). The effect must have been that of an immense and industrial-scale celebration of the Eucharist. Moreover, the whole event was broadcasted over the radio. At the exact same time when the spectators in the stadium said the Apostles' Creed (the 'I believe'), all church bells of Belgium tolled. This typically modernist event did, however, also feature many recognizable elements from traditional popular piety. Local Flemish devotional customs were incorporated in the show: the Halle cult of the Holy Virgin Mary and the penitential procession of Veurne.

These two events clearly show that it was possible, up to a certain degree, to create new devotional performances that somehow succeeded in the attempt of a radical modernization - and a substantial increase in scale - of liturgy's traditional means. However, these large-scale successes were quite rare. Catholic choral drama proved to be short-lived. It did not really survive the end of the 1930s and did not reappear after the Second World War.

Let me conclude with an anecdote. From the French critic Robert Brasillach's book *Animateurs de théâtre* (1936) comes an interesting report about a French theatre group performing in Leuven (Belgium). The 'Théophiliens', led by Sorbonne professor Gustave Cohen, were an amateur ensemble who had acquired a solid reputation producing ancient and medieval plays in modern garb. At the end of a performance in Louvain, somewhere between 1933 and 1936, the bishop of *Le Miracle de Théophile* addressed the audience with the customary invitation to chant 'Te Deum laudamus'. To his surprise, the call was answered. The audience rose and was led in prayer by the actual bishop who attended the show (Brasillach 1936: 205-6).

The Leuven anecdote provides the Catholic counterpart, as it were, to the Soviet event that introduced this essay. In this event, too, a true community came into being through inherently artificial and modernist means. Such successes, however, were not easily repeatable. Provisionally - since the subject is far from exhausted by my summary overview - I would venture to conclude that the modernist reinvention of liturgy through theatrical means was an actual possibility during the interbellum period. Its impassioned creators applied an avant-garde artistic strategy to a most traditionalist problem: how to revive the efficacy of religious ritual. But the actual happening of a successful religious 'performance' remained subject to the chance of the moment.

• **The choir of 500 angels from *Credo!***

REFERENCES

Baty, Gaston (1949) *Rideau Baissé*, Paris: Bordas.

Boon, Jozef, C. Ss. R. (1937) *Spreekkoor en Massa-Tooneel: Ontwikkeling, Theorie, Praktijk*, Sint-Niklaas: Van Haver.

Brasillach, Robert (1936) *Animateurs de théâtre: Dullin, Baty, Copeau, Jouvet, Les Pitoëff*, Paris: Corrêa.

Buck-Morss, Susan (2001) *Dreamworld and Catastrophe: The Passing of Mass Utopia in East and West*, Cambridge, Massachusetts: Harvard University Press.

Bürger, Peter (1984) *Theory of the Avant-Garde*, trans. M. Shaw, Minneapolis: University of Minnesota Press, 1984.

Claudel, Paul (1957) (1919) 'La Messe là-bas', in Paul Claudel, *Œuvres Poétiques*, Paris: Gallimard.

De Maeyer, Aloïs (1933) 'Hoe wordt het spreekkoor omschreven', *Tooneelgids* 15 April: 115-116.

De Vleeschauwer, Herman [H.J.D.V.] (1931a) 'Het spreek- en bewegingskoor II', *Jong Dietschland* 13 February: 91-92.

De Vleeschauwer, Herman [H.J.D.V.] (1931b) 'Het spreek- en bewegingskoor III', *Jong Dietschland* 20 March: 172-173.

Eliot, T. S. (1972) *Selected Essays*, 3rd rev. ed., London: Faber and Faber.

Geysen, Lode (1934) *Spreekkoren*, Roermond: Romen.

Leach, Robert (1993) *Vsevolod Meyerhold*, Cambridge: Cambridge University Press.

Marx, William, ed. (2004) *Les arrière-gardes au XX siècle: L'autre face de la modernité esthétique*, Paris: PUF.

Meyerhold, V. E. (1969) *Meyerhold on Theatre*, ed. E. Braun, London: Methuen.

Pius XI (1922) *Ubi Arcano Dei Consilio: Encyclical of Pope Pius XI on the Peace of Christ in the Kingdom of Christ*, <http://www.vatican.va/holy_father/pius_xi/encyclicals/documents/hf_p-xi_enc_23121922_ubi-arcano-dei-consilio_en.html>.

Žižek, Slavoj (2003) 'Heiner Müller Out of Joint', 25 September, accessed 20 January 2008, <http://www.lacan.com/mueller.htm>.

Bearing Witness to the (In)visible

Activism and the performance of witness in Islamic orthopraxy

DOMINIKA BENNACER

HAJJ NOOR DEEN

> After 9/11 when I walked around, I was just like, okay, at every step I feel like I am being watched. It is going to be a constant performance. You may not want to go out today, but you are going to go to the grocery store. Now then the problems within the first month or two of me going out to these grocery stores, or me going to the library, or whatever it may be, is that it was actually physically unsafe. Because there were many times when I would be cornered by people, who would be screaming things at me. And I would just have to stand there and say, 'Listen, I understand what you're saying. I am just as scared as you are. And you cannot take this out on me.' But there is no reasoning in those situations. So in order to combat that very real physical danger, I would start going out only with my friends. It worked brilliantly well in my plan though, because I was studying for orals, so really I shouldn't have been out anyways.
> (Uzma Rizvi (Interview with author, 1 March 2007)

The trauma of 11 September 2001 was experienced in at least in two disparate ways. First in the tragedy of 9/11 as a singular event and second in the cumulative trauma of the discriminatory gaze that fell upon the perceived bearers of the terrorist threat: the threatening Muslim Other. The aftermath of 9/11 invariably altered the performances of identity among diasporic Muslims whether in the attempt to efface the markers of their racial and religious identity or in the enactments that aim to highlight them.[1] The disappearance (detentions and deportations) and invisibility of Muslims in America is coupled with a paradoxical hyper-visibility. In a climate where many Muslims feel like they are part of a community under siege, the performances of religious and ethnic identity are being shaped through what Diana Taylor calls the 'complicated play of looks: looking, being looked at, identification and mimicry' (1997: 25).

This article is framed by a larger project that investigates the traumas inflicted by the US administration's domestic 'war on terror' by tracing the trajectory of the post-9/11 immigrant experience: from special registration, racial

[1] Interviews I conducted in the immediate aftermath of 9/11 revealed various reactions, from diverse forms of resistance to intensified efforts at assimilation. One Moroccan I interviewed, a man in his late 30s, stated that after 9/11 he put blond highlights in his hair and even began to learn Spanish in order to pass for a Latino. Many Muslims began to reexamine their faith and became more devout as a result. Others, who consider themselves secular Muslims, chose strategically to identify as 'Muslim' in an endeavour to contest the discrimination perpetrated against Muslims.

Performance Research 13(3), pp.64-76 © Taylor & Francis Ltd 2008
DOI: 10.1080/13528160902819331

profiling, discrimination, to detentions and deportations of Muslim, Arab and South Asian immigrants. In this essay I will consider the role of artists and activists, both within and outside the Muslim migrant communities, and the potential and limits of bearing witness in the post-9/11 security panic.

I will begin by examining the religious and legal significance of witness-bearing within Islam and the various performative and embodied forms that religious witnessing takes within traditional Islamic contexts. Here, I will pay particular attention to the relationship between the state of conscious awareness that religious witnessing mandates and the linguistic and bodily expressions of that awareness. The daily recitation of the Islamic creed, the *Shahadah*, from the Arabic verb *šahida* 'to testify', is a declarative articulation of the belief and consciousness of the oneness of God and the acceptance of the prophethood of Muhammad. I intend to consider the performative aspects of religious witnessing and the ways in which these quotidian practices act at once as the holder and vehicle of testimony.

Extrapolating from the insights afforded by the juridical and religious formulation of Islamic witnessing, in the second part of this essay I will explore the possible implications that this understanding might have on contemporary secular discourses of witnessing and testimony. By considering the work of the Visible Collective, I will examine the role of the artist and activist as witness, specifically in the context of the disappearance of Muslims, Arabs and South Asians amidst the US security panic in the immediate aftermath of 9/11.

THE STATUS OF WITNESS IN ISLAMIC JURISPRUDENCE

Within the Islamic worldview, the notion of witnessing has various dimensions and assumes diverse and nuanced meanings. Broadly speaking, the act of witnessing can be divided into two basic categories: religious and legal. In Arabic, both the religious and juridical meanings of *shahadah* simultaneously connote the action or act of testifying and the testimony borne through this very act. In legal terms, the testimony or statement (*shahadah*) of a witness (*shahid*) is a declaration on a legal claim in favour of a second person against a third. The testimony must be based on accurate knowledge of the state of affairs pertinent to the case and be made before a judge in a prescribed manner.

Traditionally, Muslim judiciary protocol does not recognize documentary evidence but only the oral evidence of eyewitnesses. In legal terminology the word 'evidence' (*bayyina*) denotes the highest form of proof - that established by oral testimony - but from the classical era onwards it came to be applied to not only to the act of giving testimony but also to the witnesses themselves (Brunschvig 1997). Thus the performative utterance and the embodied presence of the witness - if felicitous - constitutes evidence in itself.

The witness' court testimony is introduced by the words: 'I testify (*ashhadu*), concerning the rights of others' (Peters 1997: 207). This linguistic performative is an absolute requisite for the testimony to be felicitous. Statements not preceded by the formula 'I bear witness (*ashhadu*)' and commenced with other formulations such as 'I know with certainty' are inadmissible as evidence. The felicity of this linguistic performative and its legal agency can only be realized when certain requirements are in place.

Although various Islamic schools differ on points of detail of the legal status of a witness, there are some common requisites. Testimony is accepted only from an accountable person (*mukallaf*). That is, one who is pubescent, sane and has received (and accepted) the message of Islam. Furthermore, the person testifying must be upright in his behaviour and action, free from bias and unable to benefit from his testimony. Previous punishment for slander would cause a witness to be disqualified from testifying.

In cases involving property, or transactions dealing with property, the testimony of two men or two women and a man is required; alternately the testimony of a male witness along with the oath of the plaintiff is accepted. In cases that do not concern property, the testimony of two male witnesses is generally required. However, the Hanafi school holds that two women and a man may testify for marriage (Ibn al-Naqīb et al. 1994). Women's testimony of things that men do not generally see, such as childbirth, is admissible and does not require the corroboration of male testimony.[2] The requirements in cases concerning fornication or sodomy are particularly rigorous and require the accounts of four male eyewitnesses 'who testify that they have seen the offender insert the head of the penis into her vagina' (1994: 638).

The witness must have accurate knowledge of the facts to which he testifies and must have perceived them with his own eyes and ears. Generally speaking, one can only testify to what one has seen or heard, however, with regard to certain facts it is permissible to testify on the strength of public knowledge of facts without actually having witnessed an event. The importance of oral testification of what one has *seen* is underscored by the stipulation in Islamic law that the testimony of a blind person is only accepted about events witnessed before he became blind but not about events witnessed thereafter. There is a clear distinction drawn between statements pertaining to what a blind person heard, and these are admissible as evidence.

The taking and giving of legal evidence is a collective duty. However, in cases where there is only one witness present at the scene of an incident, it becomes absolutely obligatory for that witness to testify. If the concerned party requests it, it is considered unlawful to conceal testimony: 'Let not witnesses withhold their evidence when it is demanded of them', and 'Conceal not your testimony, for whoever conceals his testimony is an offender' (Qur'an 2: 282). In cases where only a solitary witness is present at an event, the witness may not accept payment for his testimony. Only in cases where the witness is not personally obligated to testify is a fee permissible. The requisition of testimony is incumbent upon the party requiring it, for the delivery of evidence is the right of the party but rests on the condition of its requisition. The only possible exception to the obligation of bringing forth testimony occurs when the observed transgression is against God alone (*hakk Allah*, right of God) and does not injure the rights of others (*hakk ademi*, right of humankind). In this case, it is at the discretion of the witness as to whether to bring the offender before the judge (kadi), yet it is considered more meritorious to remain silent, cloaking the misdeed of another.

FROM SHAHADAH TO MUSHAHADA: ON THE RELIGIOUS ASPECTS OF WITNESS-BEARING IN ISLAM

The extra-legal dimension of witnessing is of paramount importance in the life of every Muslim. The *Shahadah*, or Islamic creed, *['ašhadu 'an] la ilaha illa-llah, wa ['ašhadu 'anna] muhammadan rasulu-llah* is commonly translated: '[I testify that] there is no god but God (Allah), and [I testify that] Muhammad is the messenger of God.' This testimony of faith is the first utterance heard by an infant upon entering the world, as it is prescribed for the father to whisper the call to prayer (*adhan*), which includes the *Shahadah*, into the right ear of the child upon birth. It is also the last pronouncement made by a Muslim before dying and if the person is incapable of speaking, the *Shahadah* is recited into her ear.[3]

First of the five pillars, the *Shahadah* constitutes the core aspect of Islam and is considered the 'purest expression of the essence of Muslim identity' (Ramadan 2004: 79). It is the affirmation and testification of the faith in the oneness of God (*tawhid*) and the prophethood of Muhammad. The declaration of *Shahadah* not only testifies to a belief one might hold; it is the attestation of a reality that one has seen and recognized as true. The act of seeing could be a literal one, as in the case of eye witnessing such as that testified to in a legal context. It is also to be interpreted more metaphorically as an act of recognition or perception of a reality. This

[2] In such cases, however, the same male-to-female ratio (1:2) still follows. That is, the testimony mandated is either that of two men, two women and a man, or four women.

[3] The ubiquity of the *Shahadah* in the life of every Muslim, along with the daily acts of bearing witness to the unity of God and the prophethood of Muhammad, is effectively described in a statement by the Muslim American Society: The liturgy of Islam prescribes that the words of the *shahadah* be the first words pronounced by the Muslim upon waking in the morning and the last before going to sleep; upon conclusion of every *wudu'* (ablution) or *ghusl* (bath); in every *rak'ah* (prayer unit) during *qu'ud* (sitting on one's legs) as part of the prayer recited at that stage of *salat* (worship). The Muslim who performs his five daily rites would thus have occasion to recite the *shahadah* fourteen times a day. The *shahadah* is also recited by Muslims non-liturgically, whenever the occasion calls for it. It is used as an opener in speeches and letters, prefaces and introductions, as well as intermittently in any conversation as a means of punctuation, exclamation, or an expression of surprise, bewilderment, or reassurance. In the Muslim's mind, the notable states of consciousness are all associated with the presence of God and the subject's awareness of that presence, and therefore the *shahadah* is a suitable accompaniment. In most Muslim homes, the *shahadah* is present in beautiful Arabic calligraphy in every room, and sometimes on every wall.' (Muslim American Society 2007)

insight - the internal sight perceived by the eye of the heart and mind - when articulated through the recitation of the *Shahadah*, in turn, comes to constitute evidence of that reality. The performative utterance of *Shahadah* simultaneously acts as an indication and substantiation. It punctures the skin of *dunya*, which in Islamic terminology designates this world and its earthly concerns and possessions, and points to the more spiritual realms of existence. The linguistic and bodily expression of conscious awareness acts at once as the embodiment and vehicle of *taqwa*, or God-consciousness. The praxes associated with this consciousness, such as call to prayer (*adhan*), profession of faith and prayer (*ṣalat*), not only affirm the cognizance of *taqwa*, but act as a testimony in another sense, as a living memorial and memorializing practice. These performative and embodied testimonials attest to a certain understanding while simultaneously calling to *re*member and once more bear witness to this knowledge. First and foremost, these ritual practices perform the role of reminder to oneself. They can, however, also function on a collective level, as is most evident in the example of the call to prayer. Here the performative acknowledgement of the conscious awareness of God is at once an attestation, a reminder and a call to the active confirmation of this belief through the act of prayer.

The devotional practice of *dhikr*, variously translated from the Arabic as 'pronouncement, invocation, remembrance', is an Islamic practice that focuses on the remembrance of God. Most often, *dhikr* is associated with Sufi practice whose goal is the achievement of spiritual perfection. Various practices such as recitations of the names of Allah, supplications and recitations of the Qur'an are subsumed under the practice of *dhikr*. Although originating with the recitation of the Qur'an and other religious writings, *dhikr* gradually evolved into certain formulas including: *la ilaha illa 'llah* (there is no god but God); *Allahu akbar* (God is greatest); *al-hamdu li'llah* (praise be to God); *aṣtaghfiru 'llah* (I ask God's forgiveness). While a mechanical or mindless practice of *dhikr* is highly criticized and warned against, conscientious practice of *dhikr* is considered extremely meritorious. A hadith[4] by Bayhaqi (994-1066 C.E.) relates that the Prophet said, 'To invoke Allah Most High (*dhikr*) with people after the dawn prayer until sunrise is more beloved to me than this world and all that it contains, and to invoke Allah Most High with people after the midafternoon prayer until sunset is more beloved to me than this world and all that it contains' (cited in Ibn al-Naqīb et al. 1994: 896). Discussing the importance of holding one's the tongue and carefully weighing speech, the Prophet said, 'Do not speak much without mentioning Allah (*dhikr*), for too much speech without mentioning Allah hardens the heart, and the hard-hearted are the farthest of all people from Allah Most High', and 'All of human beings' words count against him and not for him, except commanding the right, forbidding the wrong and the mention of Allah Most High (*dhikr*)' (1994: 730). The importance of the linguistic and bodily invocations of God are further emphasized by the commentary of Ibn al-Naqīb (1302-68 C.E.):

> Perform the remembrance of Allah (*dhikr*) silently and aloud, in a group and when alone, for Allah Most High says, 'Remember Me: I will remember you' (Qur'an 2: 152). It is sufficient as to its worth that Allah is remembering you as long as you are remembering Him. Give frequent utterance of the axiom of Islam *'La ilaha ill Allah'* (There is no god but Allah), for it is the greatest invocation (*dhikr*), as is mentioned in the hadith, 'The best thing I or any of the prophets before me have said is "La ilaha ill Allah."' (1994: 804)

In this context, the linguistic utterance of the invocation of God (*dhikr*) is the articulation of a testimony in the double sense of both the attestation and remembrance.

Islamic orthopraxy strives for an absolute alignment between the state of consciousness and the bodily practices of an individual. In the Muslim tradition intention carries a substantial weight, often exceeding that of end result of an

4 Hadith are oral traditions relating the words and deeds of the Prophet Muhammad.

action. Al-Zarnuji (d. 1223 C.E.)[5] authenticates this point by quoting the words of the Prophet: 'Deeds [are measured] by their intentions' (2001: 5). Concomitantly, presence of mind while performing the prayer is of utmost significance:

> He whose heart is veiled by inattention is veiled from Allah, not apprehending or contemplating Him, but oblivious of [Him to whom he is speaking], merely moving his tongue out of habit. How far this is from what is meant by prayer, which has been established to polish the heart, renew one's remembrance of Allah Mighty and Majestic and to deepen the ties of faith in Him. As for bowing and prostrating, the point of them is certainly veneration, for if not, nothing remains but movements of the spine and head. (Ibn al-Naqib et al. 1994: 901)

Hasan al-Basri (642-728 C.E.) is quoted as saying: 'Every prayer performed without presence of heart is closer to deserving punishment' (1994: 901). Even if the prayer, the bodily movements along with enunciations which accompany them, were learned by rote, the memory of the body does not merely constitute a sophisticated system of mnemonics. The perfunctory movement of the tongue along with the evacuated *memoria technica* of bodily gestures must be transcended. The presence of mind during prayer is both the aspiration of the supplicant and a condition on which the acceptance of this act of worship is based.

The consensus among Muslim scholars is that on the path of the perfection of faith (*ihsan*) there are three spiritual stations available to a practitioner in worship. The first station, and the only obligatory one for the validation and acceptance of worship, is to perform acts of worship, by observing all of their conditions and integrals. The second station is the execution of the first stage while 'being immersed in the sea of Gnostic inspiration (*mukashafa*) until it is as if the worshipper actually beholds Allah most High, this being the contemplative spiritual vision (*mushahada*)' (1994: 815). The third and final station requires the fulfillment of two previously described stations, performed with the awareness that one is seen by Allah. This final stage is called the station of vigilance (*muraqaba*). The significance of testimony in the literal sense of eye-witnessing and bearing witness to that which one beholds with the eye returns here in an extra-legal context. For the very definition of ihsan is premised on seeing and being seen: '[The perfection of faith] is to adore Allah as if you see Him, and if you see Him not, He nevertheless sees you' (1994: 814). Although the 'as if' of the preceeding injunction might indicate that it is to be understood as a mere simile, the experience of *mushahada* as described in the second station of *ihsan*, as a direct experience of a spiritual vision, thwarts this interpretation. The experience of *mushahada* is understood not to be available to everyone but only to the elect. This injunction is further clarified in the commentary made by Ibn al-Naqi¯b, who emphasizes the importance of worshipping as if one is seen:

> *And if you see Him not, He nevertheless sees you* means that if one is not as if beholding Him in worship, but oblivious to this contemplation, one should nevertheless persist in excellence of performance and imagine oneself before Allah Most High and that He is looking at one's innermost being and outward self, to thereby attain to the basis of perfection (1994: 815, original emphasis).

This exegesis indicates the profundity of acting as if one is always being seen but also points to the possibility of imagining an inner performance of the self.

The notion of direct knowledge as experienced through *mushahada* is further elaborated by al-Ghazali (1058-1111 C.E.) who distinguishes between three levels of knowledge. The first is the faith of the ordinary people (*'awamm*), which is acquired by imitative acceptance from people believed to be truthful. The second stage of knowledge is the faith of the theologians (*mutakallimun*), which is considered to contain an element of proof (al-Ghazali et al.1995: 237). The third level is the faith of the saints (*siddiqun*) who, 'through "witnessing" (*mushahada*), experience God at first hand, and whose knowledge is hence beyond doubt. These three levels can be compared

[5] Burhan al-Din, or Burhan al-Islam al-Zarnuji, also spelled az-Zarnuji (d. 602 A.H./1223 C.E.), was a Muslim scholar and the author of *Ta'lim al-Muta'allim-Tariq at-Ta'-allum* (Instruction of the Student: The Method of Learning). His agnomen, Burhan al-Din or Burhan al-Islam al-Zarnuji, can be translated as 'proof of din' (religion, way of life) and 'proof of Islam', respectively.

to hearing a man in a house, then hearing his voice and hence deducing his presence and, finally, seeing him face-to-face' (1995: 237). Here, again, there is an emphasis on witnessing as direct perception: seeing with one's own eyes. In another context, al-Ghazali elaborates on the idea of direct witnessing: 'The meanings of the [divine attributes] are to be known through unveiling (*mukashafa*) and witnessing (*mushahada*), so that their realities become clear to [the mystics] through a proof of which all error is impossible' (1995: lxvi). The witnessing referenced here is clearly more concerned with the actual perception of the divine reality. Yet, it is important to remember that this perception is inextricably bound to the linguistic and embodied practices of Islam. That is to say, the conscious awareness of God is, in part, a direct result of the religious practices enacted by a Muslim. For it is within the context of the performative dimension of Islam that mushahada finds a structure and space within which it can occur and recur. The structure of Islamic praxis allows for a simultaneity of witnessing in the sense of perceiving and beholding a deeper layer of reality alongside a coextensive testimony, which bears witness to a conscious awareness of that which is perceived by the heart and mind.

There is yet another sense in which al-Ghazali conceives of the act of witnessing. In his work *Breaking the Two Desires*, he speaks of the *shahid* (witness), as the youth whose beauty bears witness to the beauty of the Divine (1995: 175). This kind of witnessing does not require a conscious awareness on the part of the one bearing witness but rests on the condition of the recognition of the reality being evidenced on the part of the observer. Witnessing in this sense can certainly be augmented to encompass other sentient beings as well as nature.

SHAHID(A): WITNESSING AND MARTYRDOM

Another usage of the Arabic term shahid is tied to its signification of martyr. This twofold meaning of *shahid*, which binds witnessing to martyrdom, is also present in the etymology of the English word 'martyr', which can be traced back through the Latin *martyr* to the Greek *martis*, and signifies a 'witness' and more specifically has evolved to designate believers who suffered death rather than compromise the faith. Some scholars argue that the development of Arabic term *shahid* to the meaning of martyr took place under Christian influence, analogous to the Syriac *sahda* and the aforementioned Greek *martis* (Houtsma 1987). Whatever the sources of influence may have been, it is certain that the meaning of *shahid* connoting martyr is not found in the Qur'an. The concept of martyrdom itself can be discerned in the Qur'an but never in connection with the word *shahid* and is always referenced through circumlocution: 'If ye be slain or die on the path of God, then pardon from God and mercy is better than what ye have amassed' (3: 151) and 'Consider not those slain on God's path to be dead, nay, alive with God; they are cared for' (3: 161).

However divergent from the root meaning of *shahid*, the connection between witnessing and martyrdom was nevertheless firmly established and has subsequently developed quite elaborately. What is sometimes characterized as an ultimate extension of the attestation of faith at the core of *shahadah*, martyrdom in the cause of Islam is understood by some to be the supreme manner of affirming one's faith. The boundedness of this essay does not allow for a comprehensive examination of the literature that has developed around Islamic notions of martyrdom. However, I would like to mark the contours of the various types of martyrs acknowledged in the Muslim tradition. The notion of what constitutes a martyr is highly contentious. Nonetheless, one can discern two generally acknowledged categories of martyrs differentiated largely on the basis of the burial rites accorded to them after death.

The first type of martyrs are *shuhada' al-ma'raka*, the so-called 'battlefield martyrs'. Special burial rites are bestowed upon this group and include the abstention from the customary rite of purification (*ghusl*). Furthermore, the battlefield martyrs are buried in their clothes. This practice is based on the belief that the martyr's bloodstained clothes will constitute proof of his martyr status of the Day of Judgment. It is disputed whether one should pray over the dead martyrs. The two dominant arguments against praying over martyrs come from the Shafi'is and Malikis. The first contends that martyrs are alive, while prayers are held only for the dead. The second opinion holds that prayers are intended to intercede on behalf of the dead person, but because the martyr has been cleansed of all sins, no such intercession is necessary. Both traditions are equally reliable and hence either one of the practices can be followed. However, the question of who is actually included in the category of battlefield martyrs is more difficult to settle. Generally speaking, there is considerable agreement that a martyr is one who goes into battle in order to further God's religion and in anticipation of His reward (*ihtisab*) and dies as a direct and immediate result of the wounds received. The meaning of the latter is often disputed but is generally understood to mean that the martyr must die before having the chance to eat, drink, sleep, receive medical treatment, be moved away from the site of battle or dictate a testament. All of these stipulations possibly indicate that the martyr must not have time to make the necessary preparations for death, that is, ones not made in advance of venturing into battle. The death of the battlefield martyr is often perceived as the noblest way to die, 'for in his willingness to lay down his life for a higher cause the believer overcomes the most basic instincts, fear of death' (Kolberg 1997: 205).

The second general category of martyrs is designated as the 'martyrs of the next world only' (*shuhada' al-akhira*). These martyrs are not granted special burial rites. This category is quite varied and includes: battlefield martyrs whose death was not a direct and immediate result of their wounds, those killed by *bughat* (rebels) or irregular warfare, those who die while defending themselves, their families or property against brigands or highway robbers. A sub-category of this group incorporates those persons who die violently or prematurely and include: non-Muslims murdered in the service of God, those killed for their beliefs, those who die through disease or accident. Of the latter subgroup, early collections of hadith specifically indicate those killed by the plague, those who drown, die in a fire, women who die in childbirth and so forth. 'Martyrs of love' (*shuhada' al-hubb*) according to the Prophetic tradition are those who love, remain chaste, conceal their secret and die. 'Martyrs who died far from home' (*shuhada' al-ghurba*), in turn, are those who leave their homes in order to preserve their religion in times of persecution and die in a foreign land. This list is by no means exhaustive. Before moving on, however, I will mention one final category of martyrs, as they connect the martyr back to the witness in the gnostic sense of the word. The 'living martyrs' are those who have joined the 'greater *jihad*' (struggle of the soul) and struggle with their *nafs* (self, false-ego). According to Abu al-Rahman al-Sulami (d. 1021 C.E.) the 'battlefield martyr is a *shahid* only externally (*fi 'l-zahir*); the true martyr (*fi 'hakika*) is he whose *nafs* has been slain while he continues to live in accordance with the Sufi rules' (cited in Kohlberg 1997: 206).

Martyrdom can by no means be considered the ultimate witnessing within Islam. The final and unequalled witnessing is reserved to the realm of the divine. *Al-Shahid*, the omniscient witness, constitutes one of the ninety-nine names and attributes of Allah and designates 'the One who nothing is absent from Him'. Beyond the sense of witnessing and seeing one's external and internal performances touched upon earlier, God is considered the ultimate witness on the Day of Judgment.

Extrapolating from the discussion of Islamic notions of *shahadah*, or witnessing, one can discern certain threads that run through the various dimensions of witnessing. A close examination of the etymology of the word also reveals correlation between the various uses of the term and the implicit meanings at its root. The first connotation of the verbal noun *shahida* is 'to be present', as opposed to *ghaba* 'to be absent.' Presence is the condition of witnessing, one must be present both literally and figuratively - have the presence of mind - for the act of witnessing to take place. This is closely aligned with the injunctions to conscious awareness, attentiveness and vigilance that are essential to witness-bearing. The second meaning is to 'see with one's own eyes, be witness (of an event)'. Both the legal dimension of eye-witnessing and the mystic experience of *mushahada* are figured in this definition. The third etymological aspect of the meaning of *shahadah* is to 'bear witness to that which one has seen'. Subsequently, the fourth meaning that follows is to 'attest to or certify' something. These meanings find resonance in the legal testifying by an eyewitness as well as the linguistic and bodily performances of the witness that testify to conscious awareness of Allah.

Thus far I have limited myself to the discussion of the various aspects of witness bearing within the Islamic tradition, which despite an obvious social dimension are not explicitly political and, with the exception of martyrdom, are not overtly traumatic. Now I would like to turn to the more visibly social dimension of witnessing in Islam with more obvious political overtones. The element of social activism is prominent in Islam. The call to stand against all forms of political and social oppression is expressed unequivocally in the Qur'an:

> O you who believe! Stand out firmly for justice, as witnesses to Allah, even though it be against yourselves, or your parents, or your kindred; and whether it be against rich or poor, for Allah is nearer to both [than you are]. (4: 135)

Here, I would like to examine the various forms of activist witnessing that have emerged both within and outside the Muslim community as a direct result of the US administration's 'domestic war on terror'. In the following discussion I will distinguish between two forms of witnessing. First, I will look at what has been conventionally recognized as activism and activist art as a form of bearing witness, both in the sense of making present and visible as well as memorializing practices. Secondly, I am interested in the ways in which quotidian performances of identity bear witness to individual and collective identity and in themselves constitute a mode of resistance. Before addressing the notion of activism as witnessing, however, I would briefly like to discuss a very different notion of the all-seeing eye as well as the logic of the optics of the US nervous system.

DANGEROUS SEEING AND PERCEPTICIDE

> In the murk, an eye watching, an eye knowing. Here you can't trust anyone. There's always one who knows. Paranoia as social theory. Paranoia as social practice. Michael Taussig (1992: 21)

The advertising campaigns and cryptograms, which have become ubiquitous in the United States after 9/11, attempt to modulate the nation's fear and threat level while simultaneously alerting its citizens to their own responsibility of performing a policing function. While in an unoriginal but cost-effective Foucauldian twist of its very own, the current US administration[6] is deputizing its own citizenry to locate the bearers of threat - with sixteen million eyes watching in New York City alone - the Department of Homeland Security is the apparatus whose costly gaze functions as the sanctified controller of the flow of immigrant bodies, particularly those it marks as a terrorist threat.

[6] This essay was written in January of 2008 during the Bush administration

If in his 1992 work *The Nervous System* Michael Taussig is able to characterize with uncanny accuracy our current reality as a chronic state of emergency, in which the state 'maintains the irregular rhythm of numbing and shock that constitutes the apparent normality of the abnormal created by the state of emergency' (13), this is precisely because the modus operandi of the Bush regime draws from a long-established tradition of state-sponsored terror. Although idiosyncratic in its own way, Bush's 'war on terror' deploys time-honored tactics of previous wars of silencing and state-produced terror. Nationally this lineage can be traced back through McCarthyism, World War 2 internment of

Japanese Americans, World War 1 incarceration of German Americans, 1919 detention of immigrants in the Anarchist bomb scare, all the way back to the genocide of the native peoples of North America. The disquieting consonances in the orchestration of terror with those instantiated by the Deutsches Reich, Argentina's 'Dirty War' or the Communist Regimes with their perpetual 'state of emergency', to name merely a few, resonate with and haunt the ever-present and inescapable spectre of the today's 'war on terror'. While the popular and pervasive comparisons of all forms of fascism to its apotheosis in the Nazi regime may obliterate the nuance of a more subtle analysis, the moment of the Fürer's revocation of citizenship status of naturalized German Jews, shortly preceding their 'forced immigration' and eventual extermination, is portentously congruent to the current treatment of immigrant populations. In the present moment, when the hostility towards (newer) immigrants in the United States is intensifying, when the 'undocumented', 'out-of-status' and the 'illegal' are denied human rights, and *naturalized* citizens are under constant threat of *denaturalization*, I read this reiteration of state-performed terror as particularly ominous. I propose that in the process 'naturalization' the nation-state imagines that it is humanizing and thereby endowing the newly arrived and previously purged and quarantined subject with human rights, that can with equal ease be revoked as the subject is stripped of the selfsame rights through the acts of denaturalization, dehumanization, detention and deportation and the denial of civil and human rights.

As Sheikh Hamza Yusuf (2004) has pointed out, it is impossible to wage a war on an abstract noun. Yet the current administration's strategic disinformation campaign, whose efficacy is grounded largely in a long-standing tradition of orientalism and its endless reiterations, has been to a great extent successful in perpetuating misconceptions of Islam and Muslim culture by effectively conflating the terrorist threat, misogyny and cultural atavism with all things Muslim, Arab and Southeast Asian. The repercussions of this Manichean discourse are innumerable and range from JetBlue's infamous coercion of Iraqi-American anti-war activist Raed Jarrar to remove his t-shirt with Arabic script before boarding his flight[7] to the current attacks on the opening of the Khalil Gibran International Academy, an Arabic language secondary school in Brooklyn.[8]

Originally coined by Argentinean psychoanalyst Juan Carlos Kusnetzoff to describe the military's attack on the perceptual organs of the population, the term percepticide has been elaborated by Diana Taylor (1997). One of the by-products of percepticide, Taylor argues, is that the interpretive activity of the population is reduced to an echo of the official word. While percepticide is an attack waged on the mind along with all the senses, the visual field seems to surface as one of the primary battlefields of the system at war with itself. Taussig writes:

> This understanding requires the knowing how to stand in an atmosphere whipping back and forth between clarity and opacity, seeing both ways at once. This is what I call the optics of The Nervous

7 On 17 August 2006, Raed Jarrar, a California-based anti-war activist and US permanent resident, attempted to board a JetBlue Airways flight departing JFK International Airport in New York while wearing a *We Will Not Be Silent* t-shirt written in Arabic and English. He was approached by two security officers who informed him that he had to remove the t-shirt before being admitted on board.

8 In a 24 April 2007 New York Sun article entitled 'A Madrassa Grows in Brooklyn', Daniel Pipes 'strongly oppose[s] the KGIA … because Arabic-language instruction is inevitably laden with pan-Arabist and Islamist baggage'. He goes on to argue: 'Also, learning Arabic in of itself promotes an Islamic outlook, as James Coffman showed in 1995, looking at evidence from Algeria. Comparing students taught in French and in Arabic, he found that "Arabized students show decidedly greater support for the Islamist movement and greater mistrust of the West"'. Those Arabized students, he notes, more readily believed in 'the infiltration into Algeria of Israeli women spies infected with AIDS … the mass conversion to Islam by millions of Americans, and other Islamist nonsense'.

• **Fig 1. *Nahnu Wahaad, but really are we one?*, Visible Collective/Anandaroop Roy, Naeem Mohaiemen, Queens Museum of Art, New York, 2004. Light box installation of names of detainees in the immediate aftermath of September 2001, from Migration Policy Institute database. Red lettering spells the arabic *Nahnu Wahaad* (we are one).**

System, and while much of this is conveyed, in a typically oblique manner, in the notion of the normality of the abnormal, and particularly in the normality of the state of emergency, what needs pondering - and this is our advantage, today, in this venue, with *our* terror-talk which automatically imposes a framing and a distancing-effect - is the violent and unexpected ruptures of consciousness that such a situation carries. This is not so much a psychological as a social and cultural configuration, and it goes to the heart of what is politically crucial in the notion of terror as usual. (1992: 17)

COLLECTIVE WITNESSING AND THE ARCHIVE

Human rights activist Naeem Mohaiemen, who has spent the last several years travelling and lecturing extensively about detentions of Muslims in America, speaking in a moment of exhaustion said: 'If another person asks me how many people were detained after 9/11, I'm going to have a *nervous* breakdown' (Telephone conversation with author, 15 March 2007. Emphasis mine). Is a nervous breakdown an occupational hazard of those trying to break down the Nervous System?

The Visible Collective, founded and directed by Mohaiemen, is a collective of artists and activists whose projects 'inspect hyphenated identities' and interrogate the contemporary security panic. The collective emerged from various immigrant and human rights advocacy groups that converged with the anti-war movements. The Visible Collective's early work engaged in the documentation of disappearances of Muslims in the post-9/11 security panic. Mohaiemen writes:

> The majority of detainees in paranoia times are from the invisible underclass - shadow citizens who drive taxis, deliver food, clean tables, and sell fruit, coffee, and newspapers. The only time we 'see' them is when we glance at the license in the taxi partition or the vendor ID card. When detained, they cease to exist in the consciousness. (2004-6)

Exploring alternatives to street protest, Mohaiemen along with Ibrahim Quraishi answered an open call for group shows at the Queens Museum, New York, which marked the Visible Collective's first collaborative work.[9] *Disappeared in America* opened on 27 February 2005 and consisted of a walk-through multi-media installation integrating photographs, objects, soundscapes and a film trilogy. The project, along with its *Disappeared in America* website counterpart, documents the disappearance of Muslims in post-9/11 crackdowns. Performing the labour of an alternative and community-based archive, the project traces and records the detentions of Muslims in a variety of ways. From a light box installation teeming with names of detainees, the Arabic phrase *Nahnu Wahaad* (We Are One) emerges in red. The work's title questions: *Nahnu Wahaad, but are we really one?* [Fig. 1].

And Then Things Piled Up is an interactive timeline, encompassing the period from September 2001 to July 2005, which indexes pivotal figures and events connecting them thematically to one another in a 'spider-leg mosaic of clarity, complexity and maybe even obscurity'. Defying authorized discourses and archives, be they falsified or omissive, the archiving practices of the Visible Collective destabilize the control and ownership of information and memory by disrupting the organized forgetting and official erasure of America's disappeared. Individual memories drift within communities, piling the experiences of others onto one's own, forming a collective lattice of remembrance, which in turn seeps through the community's porous borders.

Originally showcased at the Queens Museum, the Visible Collective's *Fresh, Casual American Style* subverts Gap's international advertising campaign bearing the same title by surrogating the image of Sarah Jessica Parker with photographs of detainees and their family members (Figs 2-4).[10] In an interview with *The Daily Star* (Bangladesh), Mohaiemen explains that one of the show's objectives was to 'sketch the contours of an entire community that is disappearing' (2005).

Besides the discursive absences effected by the

[9] At the time of the Queens Museum opening, collective members included Shahed Amanullah, Vivek Bald, Kristofer Dan-Bergman, Toure Folkes, Donna Golden, Amy Heuer, Aziz Huq, Sarah Husain, Ron Kiley, Anjali Malhotra, Sarah Olson, Ibrahim Quraishi, Anandaroop Roy and Sehban Zaidi.

[10] Gap's *Fresh, Casual American Style* campaign is a global one, not only catering to Europe but also opening its franchises in Asia and the Middle East in recent years.

• Fig 2: Sarah Jessica Parker in the original GAP clothing advertisement

• Fig 3. ***Casual, Fresh, American Style*, Visible Collective/Kristofer Dan-Bergman, Naeem Mohaiemen, Queens Museum of Art, New York, 2004.**

administration, there lie the literal absence and absenting of bodies. The silencing and removal of witnesses, facilitated by their deportable alien status, to a large degree dislocates the possibility of talking back. The exclusion of the individual possibility of bearing witness displaces testimony into community or collective witnessing. Agamben has argued that in testimony there is something like an impossibility of bearing witness:

> The witness usually testifies in the name of justice and truth and as such his or her speech draws consistency and fullness. Yet here the value of testimony lies essentially in what it lacks; at its center it contains something that cannot be borne witness to and that discharges the survivors of authority. The 'true' witnesses, the 'complete witnesses,' are those who did not bear witness and could not bear witness. They are those who 'touched bottom:' the Muslims, the drowned. The survivors speak in their stead, by proxy, as pseudo-witnesses; they bear witness to a missing testimony. ... Whoever assumes the charge of bearing witness in their name knows that he or she must bear witness in the name of the impossibility of bearing witness. (1999: 34)

The presence of the term 'Muslim', Auschwitz jargon for the most abject among the abject (*die Muselmänner* [Muslims]), is troubling and particularly so in the context of this essay for a number of reasons. First and foremost, the aporia of imagining a figure more dejected than the concentration camp prisoner is coupled with the label attributed to this figure, which exposes the prevalent status of Muslims as integrated into camp imagery.

Agamben uses the concentration camp figure of the 'Muslim' to theorize his notion of witnessing.[11] According to Agamben, the *Muselmann* is marked by 'the moving threshold in which man passed into non-man' (47). Not only does the figure of the *Muselmann* inhabit and represent the liminal realm between life and death, but 'like the pile of corpses, the Muselmänner document the total triumph of power over the human being' (Sofsky in Agamben 1999: 48). Primo Levi's grey zone becomes evident when one takes into account how fellow inmates treated the Muselmänner: 'No one felt compassion for the Muslim, and no one felt sympathy for him either' (Ryn and Klodzinsky in Agamben 1999: 43). *Die Muselmänner* were judged not worthy of one's gaze, considered a source of 'anger and worry', thought to be 'merely useless garbage. Every group thought only of eliminating them, each in their own way' (43).

I would argue that the Muslim does constitute 'the complete witness', although in a sense contrary to the one intended by Agamben. In a chapter entitled 'The Witness', Agamben writes,

> The language of testimony is a language that no longer signifies and that, in not signifying, advances into what is without language, to the point of talking on a different insignificance - that of the complete witness, that of he who by definition cannot bear witness. (39)

While Agamben's figurative 'Muslim' constitutes a complete witness through the impossibility of bearing witness, the Islamic designation of Muslim hinges not only on the

[11] Particularly disconcerting, given Agamben's philological and etymological inclinations, is his choice of terms to designate the threshold between the human and non-human. For just moments before introducing the figure of the *Muselmann*, Agamben takes the reader through a detailed etymological analysis of the term 'Holocaust' and, rejecting its erroneous usage, refuses to participate in its reiteration: 'Not only does the term imply an unacceptable equation between crematoria and altars; it also constitutes a semantic heredity that is from its inception anti-Semitic. This is why *we* will never make use of this term' (199: 31, emphasis mine). Instead of merely substituting the label Muslim or Muselmann to allocate the threshold of movement between the human and non-human with another term, I would argue against making such designations at all. For the only way to recognize a 'Muselmann' in Auschwitz was from the outside, based on his external appearance and behaviour. According to former inmate and self-designated *Muselmann*, Jerzy Mostowski, 'In many cases, whether or not an inmate was considered a *Muselmann* depended on his appearance' (cited in Agamben 1999: 167). Because there is no possible access to the internal psychic world of the one whom Agamben would have passing into a 'non-human' state, this threshold is indeterminable.

very possibility of witnessing but is premised precisely on the performance of witness-bearing.

Nevertheless, Agamben's discussion of the liminal figure of the *Muselmann* is potentially relevant to the examination of the complex relationship between the treatment of the abject and limits of empathy in its 'recuperation'. The threshold of indistinction between 'underman' and 'overman', life and death, human and inhuman, alongside the notion of abjection among the abject, is a blurring, which is compulsory to the very possibility of the perpetration of atrocity. For it is the defacing and effacing of the Other and her humanity that create the space in which violations of erasure find room.

The critical response to *Disappeared in America* and similar projects exposes several recurring tropes. One of them is the perceived endeavour 'to humanize the faces of "disappeared" Muslims' (Desk 2005) and to 're-humanize Muslims' (Cohen 2005). The obvious absurdity of these statements makes their unchecked circulation even more disturbing. Yet they provide a glimpse into some of the necessary, if not completely conscious, processes that give assent to current discriminatory treatment of Muslims. For what dehumanizes is not only the experience of violence but precisely the rhetoric of dehumanization that makes this violence possible in the first place. In *Precarious Life*, Judith Butler delineates the Levinasian notion of the human as figured in the face, arguing that the human is indirectly affirmed in the very disjunction that makes its representation impossible: 'For representation to convey the human, ... representation must not only fail, but it must show its failure. There is something unrepresentable that we nevertheless seek to represent, and that paradox must be retained in the representation we give' (Butler 2004: 144). If, as Taylor (1997) reminds us, the invisible is not what is hidden but that which is denied, the images of detainees and their family members figured in *Fresh, Casual American Style* show us the faces we are not supposed to see. The photographs perform an indexical and iconic function, which disrupts the hegemonic field of representation; they point beyond themselves 'to a life and to a precariousness that they could not show' (Butler 2004: 150).

Looking Away and Other Acts of Omission

The system of optics effectuated by the US administration calls on American citizens to say something if they see something, while simultaneously insisting that we look the other way and close our eyes to things not intended for

• Fig 4: Casual Fresh American Style, Project Row House installation, Texas, 2005. Man in image is Mohammed Mohiuddin (Bangladesh, b. 1972). In US for medical treatment of rare blood disease, went through Special Registration; eventually was allowed to stay in US.

us to see. Seeing is dangerous in that it 'imposes a "terrific burden"' (Taylor 1997: 25). Even if we turn away, what is unseen shapes what remains in our visual field. Levi - speaking of others at another time, of which we once said and continue to say 'never again' - writes that they 'delud[ed] themselves that not seeing was a way of not knowing and that not knowing relieved them of their share of complicity or connivance' 1988: 85). Islamic notions of witnessing bring into focus the moral obligation to bear witness against and in the face of oppression. They emphasize the collective dimension of witness-bearing. Activist art projects, as those undertaken by the Visible Collective in the post-9/11 security panic, constitute a form of such witnessing. Visible Collective's *Disappeared in America* project bears witness to what is at once unsightly and invisible: the discrimination, the detentions, deportations and torture of Muslims, Arabs and South Asians - the disappearances and erasures, both figurative and literal, that take place right before our very eyes and that we refuse to see and testify to and against.

REFERENCES

Agamben, Giorgio (1999) *Remnants of Auschwitz: The Witness and the Archive*, New York: Zone Books.

Brunschvig, R. (1997) 'Bayyina', in C. E. Bosworth, E. van Donzel, W. P. Heinrichs and G. Lecompte (eds) *The Encyclopedia of Islam*, Leiden: E. J. Brill.

Butler, Judith (2004) *Precarious Life: The Powers of Mourning and Violence*, London and New York: Verso.

Cohen, Dorit (2005) 'A Glimpse of New York From a Magic Carpet', *NY Arts Magazine* (May/June).

Desk, Swm (2005) 'Disappeared: A Project by The Visible Collective', *Star Magazine* (Bangladesh), 6 May.

Disappeared In America (2007) *<http://www.disappearedinamerica.org/>*, 1 May (accessed 1 December 2008).

Al-Ghazali, and T. J. Winter (1995) *On Disciplining the Soul(Kitāb Riyāḍat Al-Nafs) and On Breaking the Two Desires (Kitāb Kasr Al-s̲h̲ahwatayn)*, vols 22 and 23, *The Revival of the Religious Sciences (Iḥyā' 'ulūm Al-Dīn)*, Cambridge: The Islamic Texts Society.

Houtsma, M. Th. (1987) *E. J. Brill's First Encyclopaedia of Islam*, 1913-1936, Leiden and New York: E. J. Brill.

Ibn al-Naqīb, Aḥmad ibn Lu'lu' and Noah Ha Mim Keller (1994) *Reliance of the Traveller: The Classic Manual of Islamic Sacred Law* ('Umdat Al-Salik), rev. ed. Evanston, Illinois: Sunna Books.

Feldman, Noah, and Hamza Yusuf, dirs (2003) *Islam and Democracy: Is a Clash of Civilizations Inevitable?* perfs Noah Feldman and Hamza Yusuf, DVD: Alhambra Productions Inc.

Kohlberg, E. (1997) 'Shahid', in C. E. Bosworth, E. van Donzel, W. P. Heinrichs and G. Lecompte (eds) *The Encyclopedia of Islam*, Leiden: E. J. Brill.

Levi, Primo (1988) *The Drowned and the Saved*. New York: Summit Books.

Mohaiemen, Naeem (2004-6) 'Disappeared in America', *<http://www.disappearedinamerica.org>* (accessed 1 December 2008).

Muslim American Society (2007) *<http://www.masnet.org/history.asp?id=321>*, (accessed 11 December).

Peters, R. (1997) 'Shahid', in C. E. Bosworth, E. van Donzel, W. P. Heinrichs and G. Lecompte (eds) *The Encyclopedia of Islam*, Leiden: E. J. Brill.Pipes, Daniel (2007) 'A Madrassa Grows in Brooklyn', New York Sun, 24 April, <http://www.nysun.com/article/53060?page_no=1> (accessed 27 April 2007).

Ramadan, Tariq (2004) *Western Muslims and the Future of Islam*, Oxford and New York: Oxford University Press.

Ryn, Zdzisław and Stanislaw Kłodziński (1987) *An der Grenze zwischen Leben und Tod. Eine Studie über die Erscheinung des 'Muselmanns' im Konzentrazionslager, Auschwitz-Hefte*, vol. 1, Weinheim and Basel: Beltz.

Sofsky, Wolfgang (1997) *The Order of Terror: The Concentration Camps*, trans. William Templer, Princeton: Princeton University Press.

Taussig, Michael T. (1992) *The Nervous System*, New York: Routledge.

Taylor, Diana (1997) *Disappearing Acts: Spectacles of Gender and Nationalism in Argentina's 'Dirty War'*, Durham, North Carolina: Duke University Press.

Yusuf, Hamza (2004) *Religion, Violence and the Modern World*: UK Tour.

Al-Zarnuji (2001) *Instruction of the Student: The Method of Learning*. Trans. G. E. Grunebaum and Theodora M. Abel. Second Edition ed. Chicago, I.L.: Starlatch Press.

Equal are the Maasai and God

MALCOLM FLOYD

After a millennium of movement down the Rift Valley of East Africa the Maasai, a group of peoples related by language and cultural practice, settled in the area of what is now northern Kenya to northern Tanzania.[1] My acquaintance began when I was teaching in Nairobi and was invited by a student from the Samburu section of the Maasai to stay with his family during one long vacation in 1983. Most of my early research was done with this particular family and their community. In the early 1990s, I travelled to research with other groups, including those in the areas around Dol Dol in Kenya and Arusha in Tanzania. After that period I wrote a chapter in Ralls-MacLeod's and Havery's Indigenous Religious Musics called 'Maasai Musics, Rituals and Identities' (Floyd 2000a). In this current article I shall revisit that earlier chapter and 'write across' it. It concentrated on the ways ritual reifies religion, and how ritual was extended beyond its identifiable times by the wider use of songs. My focus then was on song texts. However, I have become aware that there was also experience involved. My earlier article still clung to an archetypal Western position of being 'separated' from the research, if not 'objective', and I focused on the organization of ritual and identification of this as 'religion'. But in some cases I shared in the experience. When I talked with the Maasai about what was going on and what happened, we seemed to share a notion that these were phenomenological moments, where the experience placed itself richly and comprehensively in our consciousness.

In this article I want to explore the experience of this performance, of ways in which spirituality can be considered to be embodied and vocalized, and how one (Maasai and visitor) can come to share the experience. I shall set out something of the Maasai themselves and their singing, then consider notions of spirituality. Finally I hope to explain how my reconsidered experience worked, and indeed still works.

[1] There is a summary of this process in Floyd 1999 (135-6).

MAASAI, GOD AND RELIGION

Maasai-ness is generally considered to be a historically constructed, pastoral nomadic 'cattle culture', with both lineage- and age-based social structures, in which ritual focuses, defines and reifies cultural identity. The significance and interconnectedness of cultural signifiers is communicated through actions - biological, technical or expressive - of which ritual is a fundamental example. Such actions are in the nature of adaptive systems, resolving problems in the relationships between people and their habitat, each other and the unknown.[2]

[2] There is an elaboration of this conceptualization in the first chapter of Floyd (2000).

My starting point in 2000 was to focus on a notion of 'God', as being the starting point for discovering 'religion', and drew on some of the literature explaining Maasai views and constructions of 'God'. I wrote then:

> For the Maasai, *Ɛnkai* (God) [also spelt *Ɛngai*] exists as mythical bi-polar historical focus, and as a target for prayers, which exist mostly in the form of pleas from wives for children. The mythical focus is apparent in a number of stories and other items of oral literature ... For example in *Ɛneimua enkiteng'*

Performance Research 13(3), pp.77-88 © Taylor & Francis Ltd 2008
DOI: 10.1080/13528160902819356

(the origin of cattle), God is said to have intended to give all cattle into the hands of Maasinta (mythically the first Maasai), and was only interrupted by a Dorobo house-mate of Maasinta who exclaimed out loud in his shock, so making God think Maasinta had enough cattle (Kipury 1983: 30-1). (Floyd 2000a: 85)[3]

[3] Passages from 'Maasai Musics, Rituals and Identities' are given in boxes.

The Maasai have used this particular myth to justify both their raids on the cattle of others and their subjugation of the Dorobo, a small community of hunters and ironworkers. On one hand this seems to be simply a *post hoc* justification for straightforward theft and oppression, but it can also be read as placing *Enkai* in a position of agency for the Maasai, here in particular for the warriors. It also places *Enkai* under an obligation to the Maasai because of this initial mistake. This is, of course, something of a reverse of the Adam and Eve story where humans put themselves under an obligation to God because of 'original sin', made more emphatic by the lack of judgement in listening the serpent. The sin here is Enkai's, possibly shared with Maasinta.

The Maasai are aware of the nature of the implications of this relationship as can be seen in these two proverbs:

Erisio Ilmaasao o Enkai and God	Equal are the Maasai
Erisio olporror o Enkai with God	The age-group is equal

(Kipury 1983: 150)

This position of agency for, and obligation to, warriors is comparable to the use Enkai is put to by women, particularly those who have not been able to have children quickly enough, emphasized in the urgency for women to have sons who will become warriors in their turn. Here is an example of one such sung prayer text:

Laitoriani lai tooruaki ngura	My lord comfort me
Pamakut lpaiyani tongopang mock me	So men will not
Keto iyeu nkera	I want children
Ainyo pamayou ntolut precious gift?	Why can't I want a

(Sung by women in Maralal, Kenya, August 1984)

The bipolarity is apparent in stories such as *Enkikurrukur o nkaitin* (Thunder and the Gods), which talks of two gods, black ('humble, kind and loving') and red ('malevolent and did not care about people at all') who struggle to achieve mastery over the Maasai, and concludes:

> And so, to this day, when one hears loud thunder, it is the red god who is trying to get past the black god to wipe out the people on earth. But when the sound of the thunder is not very loud, it is the black god who is trying to prevent the red god from killing the people.
> (Kipury 1983: 29-30)

Donovan describes this bi-polarity as being contained within the Maasai concept of God itself:

> For the Maasai , there is only one God, Engai [alternative spelling], but he goes by many names. Sometimes they call him male, sometimes female [the en- prefix indicates a feminine noun]. When he is kind and propitious they call him the black God; when he is angry the red God. Sometimes they call him rain, since this is a particularly pleasing manifestation of God. But he is always the one, true God. (Donovan 1982: 42) (Floyd 2000a: 84)

While Donovan, a Catholic priest, worked hard to understand Maasai notions of 'God', he found it difficult to relate this to his notions of what 'religion' must consist of and he concludes that Enkai was separated from any organized 'religion'. He describes

> with some fascinated bewilderment the fact that they have no ancestor worship, no immortality and no burial of the dead, relying instead on hyenas to deal with corpses. (Donovan 1982: 21)

However, there is a notion of 'god' that allows the Maasai to employ him/her for particular ends as defined for particular times, circumstances and places. But this is to take a distanced and pseudo-objective tone.

In 2000 I sought to make a partial case for the existence of religion in the person of the only 'religious' figure among the Maasai, the *Laibon* (variously translated as witch-doctor, prophet, diviner, shaman, seer and ritual-leader). Certainly he is consulted to foresee the future, but it is in his ritual-leadership that his role is most fully defined: he officiates at rituals, and he names, starts and concludes age-sets. So if *Enkai* is not at the centre of Maasai religion, perhaps religion is much more clearly seen as being reified in rituals, which provide both cyclical and linear patterns of cultural construction. That notion provided a useful framework for my earlier work and allowed the accumulation of data and evidence for analysis and interpretation. However, it dealt with religion in a mechanistic way. In retrospect my patronizing of Donovan was echoed in my own approach: we were looking for a set of 'proofs' for the existence of religion. Donovan came with his pre-existing notions and found them lacking; I came with a willingness to allow anything that found a belief in God and ritual, and equated this with religion, in spite of the complexity of the Maasai notion of God and of her/his connectedness with them. It is worth spending a short time examining the construction around ritual that I made and the flexibility within it that seemed inescapable.

RITUAL

It is predominantly the age-set system that is the focus for ritual, both for men as members of age-sets and for women as attached to their husband's age-set. This bond undoubtedly has the strongest call on any individual's loyalty. I chose to become informally attached to a young warriors' age-set from my first period of sharing time with the Samburu. This worked for everyone because as an unmarried man it was the appropriate place to be, and because my father was older than any of the others it gave me a position of relative rank, which meant I could do the things I wanted to do, including asking questions, making recordings and so on, and avoid others. It also meant I was not in a competition for status with the elders, which would have proved very unwieldy and politically challenging. I make this point here because it shows the setting-up of a context of inclusion, and inclusion on favoured terms at that. This often serves to promote loyalty and commitment and encourages the incomer to take further risks in opening up to the implications of joining a new group. It provides a platform for encounters with events that can be treated as experience, rather than just data or evidence. I am not alone in this, of course, and it is, and has been, common in the work of ethnomusicologists and anthropologists for a long time.

For 'Maasai Musics, Rituals and Identities' I constructed a table that set out something of the intricate sequence of ritual through Maasai life. I argued in 2000 that this showed

> a highly structured society', with clear concepts of progression, marked by rituals, with particular and discrete elements, with the unifying feature being variations of meat-eating feasts. This emphasizes the importance of cattle, and other livestock, not only as food, but as important features of Maasai ritual life. (Floyd 2000a: 90-1)

On further consideration other emphases become known. Maasai ritual life is about experience: moments when participants are aware that something is happening or has happened and is having or has had an impact on them. The structure of ritual rotates the focus of communal and individual concentration so that each group becomes the centre of attention, young and old, men and women. This progression in ritual forms overlapping narratives: one of which is linear (that of the individual, moving from one point of recognition to another) and the other cyclical (that of the community, which might experience many focus moments across a typical year), thus both are mutually reinforced. The individual moves on and gains status and responsibility (or at least is credited for getting older); the community remains stable and reliable and frequently makes a point of remembering. Once the habit of remembering is established, its elements seem to gain the status of 'tradition'. This notion of developing while at the same time being contained provides for the Samburu a security. This, in turn, allows the safe performance of outrage and rebellion.

THE EXPERIENCE OF SONGS

During one period in northern Tanzania in 1995, I was able to have discussions about songs and rituals with Maasai students in two schools, one secondary and one primary. It should be noted, however, that ages vary in Maasai (and other) schools, so the thirteen discussants (six male, seven female) in the primary school were from 13 to 16 years old, the eight (six male, two female) in the secondary school were between 18 and 23 years old. Our discussions ranged around the relationship between ceremonies (their preferred term for ritual), songs, their thoughts and feelings and anything that might have been 'learned'. The shared completing of questionnaires was used as the stimulus for discussion.

It is interesting to revisit that material at this distance of time, and I was surprised to see the emphasis on feelings and to realize that much of this material never made it into MMRI. The primary school students all said that they enjoyed singing and liked listening to songs as well, with three saying it made them 'feel good'. As to why they liked to sing, they were agreed that it was because of enjoyment, with three boys agreeing with the statement of one that it was 'because I love a song'. They said that they felt like joining in when they recognized the 'good arrangement' of a song. This seemed to consist of it being suitably pitched and confidently started by others. However, there were two (one boy and one girl) who put forward the notion that they wanted to sing when they were 'not in a good mood'. As to what was communicated through, or learned from, songs, four of the girls thought that it was 'to behave in good ways of living', while the boys concentrated on whom they had learned *about* in learning *from* them. So for one boy that was his grandmother, for four others it was their grandfathers. There was a short discussion about times and places for singing, and it was interesting to note that evening was one of the times when they felt it was inappropriate for them to sing. The reason for the interest is that evening is a time when many Maasai do sing. Certainly I have been present at singing of the warriors, sometimes with young women and even girls, and also of women. But these 13 to 16 year olds were at an interesting stage. They were probably not yet initiated but it would not be long, and the boys in particular would have been at the stage of extreme teasing and bullying to encourage them towards the ritual with some eagerness and determination. When we moved on to discuss warriors' songs in particular, those who were prepared to discuss their reaction to these talked about thinking about their parents. One boy and one girl mentioned turning to their mother, four boys talked about their fathers. The implication was that this was a search for safety in a potentially frightening situation. As for what was frightening them, it seemed to be the quality of the sound as much as the texts.

Of the older students at the secondary school, all but one said that they enjoyed singing and that songs made them feel happy or very happy, indeed one girl said 'If I was sitting, I stand and start playing'. This playing seemed to relate to moving in response to the music as a prelude to joining in. The one dissenting voice was from a boy whose strongly held position was 'I am saved, and my religion forbids it'.

One Arusha boy said that songs made him feel both 'very happy' and 'very sorrow'. When discussing what made them feel like singing, there was an opening out from the mood enhancement or change of the younger students with comments such as: 'to enjoy myself'; 'the songs make me happy and teach me some of the way how to live in this world'; 'to teach others various things'; 'the important message in the song'; 'connection society'; and 'people can like me because of my songs and singing'. This strong moral imperative was developed in their analysis of what they had got from songs. For the boys this included: 'how to take care of cattle'; 'to be a strong man, to defend your society, to work hard and to control your family' and 'to be kind, be a strong man, to work hard, not be lazy'. The girls' approach was rather different: 'maybe what you want to do when you finish education'; 'when we finish school there will be many problems' and 'working hard is the key of success, parents must be able to care of children, love between people, behaviour'. This gender distinction remained as the discussion considered warriors' songs in particular. Two of the boys talked about them being 'good and nice, and make us not to become coward' and 'it qualify the warriors and society', while a girl said 'I think of where to get help when such a warrior was on my side'.

These older students were asked if they would write down anything else they wanted to share about their music, and in reading these it is possible to read a strong sense of identification with their community and its practices and of their understanding of how the two things relate, in the sense of being mutually efficacious. Some extracts follow:

> The Maasai is the one of the songs I like and I know very much because they cause me to mention my tradition. [male 22]

> My music mades very great things because there are songs which when sing made people happy even there are other help in teaching society what will be suppose to do and when your playing they made people be happy. [female 19]

> My music is so well so I want to say these. Our music is so beautiful that can make you feel very happy and forget all about your problem and trouble ... So I want to challenge all young men who belong to our tribe to like our tribe much than those foreign one. [male 21]

One takes the opportunity to suggest ways of sharing and supporting this rich resource:

> I advice the government to look if there is any way or solve this problem to teach this songs in school so tat other people would be attracted. Also to record our music so that to be known by everyone; to write our music in books so that the next generation can know how they are to compare with them. [male 19]

It is important to balance this again with the Christian boy who resists the apparent allure of the traditional:

> Nothing I can write because I am not traditional music. Because as I told you that I have been saved. ... Christians in my religion are not allowed to sing traditional music because if you sing you will make sin to God ... Traditions songs glorified traditions values. That is why as a Christian I don't like to sing them. [male 21]

The notion of mutually reinforcing linear and cyclical narratives, relating to the individual and the community, is apparent in these discussions. For these students songs are reserves of emotional support, social contact and integration, values and attributes for them to espouse as they grow up into adulthood. This seems to be close to notions of self-fulfilment, (of 'self-actualization' to use another term): it is the perfection of the self within a perfect community. By 'perfect' I mean 'in a condition that meets all perceived needs'.

Particular times are important and modes of behaviour are identified as ideal. The enhancing or even changing of mood is fundamental; the ultimate 'one-in-many' feels a particular way. The many are identified as friends, parents and grandparents, and it is interesting the note the role of grandparents in sharing cultural paradigms. The 'one-in-many' (or 'I-in-we') also has duties to help provide purpose. These include teaching others and working.

All of this comes from singing. Singing a few songs together, fairly frequently and around particular occasions - and for particular purposes there is an added emphasis on such singing. I want to take a little time to relate this to my own experience.

SHARING THE EXPERIENCE OF SONGS

Since childhood I have always sung in choirs. I was fully enculturated in the traditions and practices of English choral performance. But there was a challenge to my way of thinking about it and being in it when I went to work in

Kenya when I was 24. I was travelling in the north of Kenya and was present at a performance by some Samburu warriors. I was particularly intrigued by the fact that they ignored us, what I had assumed was an audience, and concentrated on themselves. They were in a circle facing inwards, the singing was for them and had consequences for them. I have often wondered what it was that affected me by that performance; now I think I identified that quality of 'I-in-we' (what I shall now call 'wi' - pronounced 'we'). There were separate elements of performance, with requirements for individuals to take the lead and sing about themselves and their exploits and then return to the group. There were also things to do with the qualities of tone that seemed to support this notion. When I started to study with the Samburu, I was encouraged to sing with them. At first I sang with the group parts, to assure my membership, and then gradually started singing a few short solo passages. This took some nerve, but in a strange way having fifteen warriors laughing at your latest attempt is strangely supportive as well as motivating. I think this is because the laughter had an innocent element rather than a sarcastic or judgemental one. They simply found it funny. But I was becoming part of 'we': becoming 'wi'.

What, then, is this quality in Maasai singing? How does it relate to notions of religion, spirituality and congregation?

MUSIC OF THE MAASAI

In 'Maasai Musics, Rituals and Identities' I wrote the following summary:

> [Maasai music] is based on individual and communal experience, and requires individual creativity within a well-established framework, and with a recognisable repertoire of appropriate motifs, to which the individual is supposed to add. Within this there appear to be three principal repertoires; for infants, for warriors and for women. The music for infants is often led by older girls and young women, and there is no gender separation of children at this age. Songs are mostly call and response, with simple response parts, and may be interspersed with riddles to encourage mental and verbal dexterity ...
>
> For Maasai men musicianship has to do with recognition and assimilation of short accompanimental motifs and progression beyond that to the creation of new motifs, which will need to be approved by the group in performance and which will then become part of the individual's repertoire, which can be drawn on as occasions demand. Musicianship also requires mastery of the rhythmic semi-pitched hyper-ventilation nkuluut technique ... The third part of male Maasai musicianship is the assimilation of melodic motifs to be used as a basis for singing one's own song, to boast of one's heroism and to encourage female admirers. Each boy has to work at these aspects, balancing individual achievement within communal performance. All these individual skills come together to create group performances, and for each performance the individual selects what to contribute.
>
> Married women create a new repertoire exclusive to them, not drawing on the songs they have sung as children or while courting. It has two main aspects:
>
> 1. Laomon: Prayers, particularly for children. This explicitly religious music is almost entirely restricted to married women.
> 2. Kagisha: Songs sung to praise warriors, and to mock their husbands ...
>
> The songs themselves provide the information and, for women, most usually consist of a solo phrase, followed or overlapped by a choral unison response. There are those considered expert, but there does appear the same requirement that all should be expert, and time is given to achieve skill, which is often focused at times when infertility and interest in men other than the husband become especially important to individuals. (Floyd 2000a: 93-4)

It was obviously clear to me that this relationship between individual and group was important. It is worth emphasizing here that it is important for both men and women. Because of the structure of Maasai society and of my part in it, most of my experience was of listening to, and singing with, the warriors. I learned to 'tune my voice' to theirs. This was never going to be possible or appropriate with women, and I did not have the same instinctive sense of identification with their singing tone. However, it is no part of my intention to demean the significance of their singing.

In 'Maasai Musics, Rituals and Identities' I spent some time analysing song texts to unearth concepts that I considered were to do with religion, ritual and so on, as described earlier in this article. Here I have taken a different approach as I have discussed the notions and the chart with Maasai in the intervening period.

Concept Analysis of Maasai Songs

	total 95		*total 34*		*total 115*	
Concept	*M no*	*%*	*F no*	*%*	*Total*	*%*
Women	19	20.0	15	44.1	32	27.8
Warriors	36	37.9	14	41.2	44	38.3
Children	**54**	**56.8**	**21**	**61.8**	**67**	**58.3**
Elders	13	13.7	8	23.5	21	18.3
Whites	13	13.7	2	5.9	14	12.2

There are many songs to do with children from both warriors and women. The warriors are often exhorting boys to get ready or chastising them for not being ready. The women are generally praying for them, as described above.

History	**45**	**47.4**	**8**	**23.5**	**50**	**43.5**
Cattle	**39**	**41.0**	**15**	**44.1**	**49**	**42.6**
Agriculture	2	2.1	1	2.9	2	1.7
Travel	**55**	**57.9**	**18**	**52.9**	**65**	**56.5**
Social. org	**45**	**47.4**	**19**	**55.9**	**57**	**49.6**
Soc.behav	**43**	**45.3**	**21**	**61.8**	**57**	**49.6**
Education	-	-	1	2.9	1	0.9

There are a fair number of songs on these elements. This would seem to fit notions that songs are about teaching and reinforcing behaviours and practices. These songs also remind us that formal education is not generally considered of much importance. There are other things more worthwhile.

Rebellion	2	2.1	7	20.6	9	7.8
Competition	2	2.1	2	5.9	2	1.7
Anger	**11**	**11.6**	**4**	**11.8**	**15**	**13.0**

While Paul Spencer and I have written about rituals of rebellion, in fact the songs say very little about it. The message then is not limited to the text, and may indeed be excluded from it. It does, of course, vary with circumstance and particular improvisation. There are a few songs about anger, however, which probably serve for most occasions.

Fear	8	8.4	9	26.5	16	13.9
Punishment	8	8.4	1	2.9	9	7.8
Criticism	12	12.6	8	23.5	18	15.7
Praise	**15**	**15.8**	**9**	**26.5**	**21**	**18.3**

The paucity of negative approaches to developing community is very interesting, although they do exist of course. Parallel with this, however, is a relatively small number of songs that praise. Songs then do not intend to motivate through their texts, at least as a general principle.

Ritual	10	10.5	2	5.9	11	9.6
Initiation	8	8.4	1	2.9	8	7.0
Religion	13	13.7	7	20.6	20	17.4
Xtianity	6	6.3	3	8.8	9	7.8

There is very little specific to these ritual or religious elements. Again, we see that songs are not *about* but *for* their circumstances, purposes, functions and uses.

Opposite sex	**51**	**53.7**	**23**	**67.6**	**65**	**56.5**
Same sex	**52**	**54.7**	**15**	**44.1**	**61**	**53.0**

There is a predictable concentration on gender. What might be less anticipated is the similarity of the concentration on people of the same sex. For example, in one song a warrior sang, 'I love ole Nakuya like I love my stomach' (319)[4]. However, this can obviously be read as a locus for developing 'wi', as the individual is, of course, gendered.

Value	**65**	**68.4**	**26**	**76.5**	**85**	**74.0**
Identity						
Indiv	**66**	**69.5**	**23**	**67.6**	**78**	**67.8**
Comm	**65**	**68.4**	**28**	**82.4**	**85**	**74.0**
Nation	**-**	**-**	**2**	**5.9**	**2**	**1.7**

4 Song numbers given refer to their number in the author's collection, which is accessible at the British Library National Sound Archive, item C810.

Here we find a remarkable focus. First on values, by which I mean articulations of things the Maasai say/sing they value. What is valued is sung about. An essential value is identity. Indeed, from this evidence it is the principal value. Predictable from what we know so far, notions of individual and community are given equal prominence in the texts. As one song goes: 'Sharing one stick ... one club ... one mother ... one father' (300). We can also hear how an individual might start to create 'wi' in a new place: 'I don't know this place because I am a stranger, so I need guidance specially on a raining night (287).

What exists very minimally is the notion of 'nation'. As always, this is dependent on particular occasions and improvisation choices. On this occasion a Member of Parliament had visited the area a few months earlier, and the women had been required to perform songs for him. Even so, there is little space in the Maasai notion of identity for an unknowable community. 'Wi' is local; it consists of 'I' and 'we'. It is more difficult to make 'we' an abstract formulation (particularly when so many are 'not-Maasai').

Development	31	32.6	7	20.6	36	31.3
Stasis	**54**	**57.8**	**24**	**70.6**	**71**	**61.7**

The notion of 'being' is continuous, but for the Maasai it has some containment. This is often part of a deliberate move to avoid moving into wider structures such as organizations, urban environments, education and notions of multi-culturalism. There are songs that comment on change, particularly where the advantages seem relatively clear of cultural disruption. However, even electricity, which it is acknowledged would bring many benefits, is also seen as problematic as it ties people to particular houses, requires money rather than trade, makes locations consistently brighter so reducing the impact of moonlit nights, and making houses lose their particular fire-lit quality, which allows for a degree of anonymity, of hidden-ness, indeed of 'wi'.

Song reference	46	48.4	11	32.4	50	43.5

Songs themselves are referred to within the songs. This is generally to get it going again, often a request to people to 'tune their voices', or to suggest a change of topic, or venue (perhaps to move outside when they have started off inside).

So we have a soundscape in which songs do not often relate specifically to particular times or circumstances but are employed to particular ends and are part of reifying the process of becoming. I expressed this in 'Maasai Musics, Rituals and Identities' thus:

> [T]here are repertoires of songs and song types that are employed widely and that have resonances with the focus of rituals. (Floyd 2000a: 97)

We see a focus on children, wanting to have them and then wanting to prepare them for full integration. There is a concentration on cultural practices and behaviours, including preparation for work. There is a concentration on a range of relationships, with same and other sex. There is much on the relationship between the individual and the group but this is not extended to wider notions such as the nation. The preference is for developing the process of being/becoming Maasai in a mainly static form. And the songs themselves are self-referential and reveal something of the process of becoming 'wi'. If my Maasai colleagues and I are right, then this state of 'wi' is more than something to be aimed for; it is an achievable state of perfection (following my earlier definition). The most admired and respected in Maasai society are those who most completely embody and exemplify this state. I argue that they become adept through singing, as music provides a temporally indeterminate space in which ritual and cultural moments and ideas can be located and identified. As many rituals do not generally involve explanation or indeed much speech of any sort, music takes on that role of exegesis and preparation. Indeed it might be described as an asynchronous liturgy. So the cultural/ritual concept is kept continuously in the mind and voice, providing the context for ritual and elaborating its meanings for the complex overlappings of the identities of individuals and the community. There are times when the music becomes the ritual/cultural moment.

I make large claims for this performance, and I want to explore that further by examining the experience. The Maasai male voice called to me. It sang itself to me, it invoked the something of what I came to recognize as 'of Maasai' in me. At least that was how it felt. On reflection, that is to do with the attractiveness of the voices to me, and the attractiveness was to do with feeling it was possible that my voice could 'fit' with theirs, indeed that I might be able to 'tune my voice' to it. As Barthes writes: 'The grain is the body in the voice as it sings' (Barthes 1977: 188). This is what I heard and felt. My trained voice never really 'worked' with popular or folk styles, as the wrong grain was in it. But here I heard something different. All descriptions of voices are subjective, but I'm interested to look back through my notebooks for this earlier period: 'it sounds so open, there's a lack of constraint and a real sense of the sound coming from deep within the body, and that's where the movement seems to come from'. Indeed, the Maasai generally dismiss the notion that their jumping (as seen on a recent BBC ident, 13 August 1986) is dancing. It is the movement that the singing requires, and thus they have an interdependent relationship. It was the jumping that took the most time for me to engage with as it was no part of my training. But being prepared to jump allows the voice to open in the way that works to create the Maasai sound. Similarly, the *nkuluut*, rhythmic hyper-ventilation, creates a sound that uses all the air from inside the body and thus the body expires - it gives all, sometimes very dramatically when a warrior falls into a catatonic state and has to be rescued by his colleagues.

Melodies are developed from motifs, but there is freedom to do that in individual ways and to construct words and music simultaneously. When I first started trying to join in, part of the amusement for the others was caused by the fact that I had so obviously been preparing them for hours beforehand and hoping the right song would come along. It was a permissible thing to do but not one that any of them would confess to, and I never heard anyone trying anything out quietly. Freedom with singing seems to come with embracing the risk and enjoying the adrenaline rush. (This, combined with oxygen deprivation, can lead to a very different performance from many others I've been in as a member of a vocal group.)

This experiential notion is recognized by some of the Maasai with whom I have discussed the idea. When looking at some of the discussions, they have said, 'It is why they are happy', 'they feel that happiness which is more than just happy, it is being Maasai' and so on. I have been calling this experience the achievement of 'wi', with some becoming adept and thus admired and respected as they (in we or they) get to the point of perfection (remembering the particular limits I am setting on this). This is taking on a spiritual dimension, entirely in line with a particular idea of congregation, and it is that spirituality I want to consider next.

SONGS AND SPIRITUALITY

I want to start this within my current context, the place where I am now reflecting on all this, and on three academics from my particular university 'congregation' writing on spirituality. Anna King tries to pick a way between essentialism and relativism, so spirituality is not always identifiable in a particular way, but neither is it just defined by the individual in the moment. There is a recognition of the dynamics of culture and mutual shaping processes and that this needs a 'language attuned to it' (1996: 350). For King this brings in an understanding of spirituality as 'the capacity of human beings to create or perhaps discover value and meaning' (350). There is also a mutuality inherent in spirituality where 'persons create and recreate transforming visions of life in the very flow and untidiness of our experience' (351). Her conclusion is that spirituality is essentially a place for transformation and change and that studying spirituality requires a methodology of metamorphosis.

One initial reaction to this is that the Maasai version seems to something of a resistance to it. Concepts of transformation and change at the macro level seem to be limited as the Maasai deny nation, development, globalization and so on. However, if we consider notions of inter-group and intra-group transformation and change, the relevance is much clearer. Indeed, if we start looking from a Maasai perspective, it is fully realized. Changes from one stage to another are highly organized, and songs are used to sing them into being. Becoming, or maintaining a state of being, Maasai requires a constant adjustment and/or reinforcement of values, practices, actions and processes. Song is the medium in which this happens. So there is transformation and change, but to particular ends that often, however and rather contrarily, are to do with retaining the status quo and an identifiable cultural-artistic pattern.

June Boyce-Tillman has developed her notion, first expressed in 1986, that performance, and music in particular, consists of values, construction, expression and materials, and that spirituality (which she has also described as 'transcendence') exists in the relationship between them seen as four overlapping circles. She then goes on to consider the qualities that result from this relationship. These include an ability to transport an audience to a different time/space dimension ('from everyday reality to another world'), flow, ecstasy, trance, mysticism, peak experiences and liminality (2007: 209). These are certainly qualities that are used in describing and even perhaps recognizing a 'spiritual' experience. I am aware that there is a danger here of attempting to universalize the experience, but the notions are so frequently encountered in discussion that they seem to have resonances in a range of contexts, and I am interested to see if they work for the Maasai.

Boyce-Tillman's four 'elements' of music are apparent. In a sense they are such core aspects that they are inescapable. Songs are sung by voices, those are the materials; songs are constructed, there is at least a sense that one part/motif/idea/phrase is followed by another, which will be either the same or different; songs are sung with some expressive intention, whether that be emotional (to help enhance happiness) or social (to sustain relationships with one's peers) etc; and finally songs are value-bearers, there are shared tonal features, there are songs to show the significance of 'wi' and so on. The qualities she proposes seem to be less securely in place. There is not a sense of 'another world'. Donovan was unable to find such a notion, and I have not found it either. This is because in having perfectable 'wi', and of that being embodied and exemplified from among themselves, they are, as quoted earlier, 'equal with God' (Kipury 1983: 150). Everything happens here, among 'wi'. There are mysterious moments, such as the fainting into catatonic states that warriors do on occasion. This seems to be related to a heightened sense of anger and the hyper-ventilation of some accompaniment types, but the 'mystery' of the moment has been described as being too angry and trying to contain it. There is not, otherwise, a tradition of ecstasy or trance. Indeed, this is all part of the Maasai 'being equal with God', as there is nowhere to be above or beyond, nothing to be imagined, all is here and now and Maasai. But there are peak moments, when singers will have sung for a long time, voices have been 'well tuned', people have improvised satisfyingly and interestingly, *nkuluut* and jumping have given the voice full body, and all have been able to join in across all parts of melody and accompaniment. This is the moment when 'wi' are the champions: 'Wi' sing our identity, our relationship, our Maasai-ness.

This might, of course, be considered a very romantic, even Romantic, notion in the sense of an appeal to the emotions, being idealistic and of creative imagination (but not in an uncontrolled form)[5]. But if we consider Vaughn's idea that Romantic Art gives the 'suggestion of a deeper and otherwise unknowable reality beyond what can be experienced' (Vaughn 1994: 11), then he is not talking about the Maasai. All depth is among us (and I am choosing to emphasize my relationship); what is known is what is known and there is

5 http://www.google.com/search?hl=en&rls=com.microsoft:en-us:IE-SearchBox&rlz=1I7RNWE&defl=en&q=define:romantic&sa=X&oi=glossary_definition&ct=title accessed 27.01.08

nothing that we want to know that cannot be known and there is nothing beyond experience.

Finally I want to draw on some of the thinking of Anthony Noble, who is interested in the notion that there is within music a type of perfection (2008). He discusses Jeremy Begbie's notion that 'it is the power of the Holy Spirit which gives "inspired" artworks what perfection they have' and John Butt's statement that; 'the very substance of music both reflects and embodies the ultimate reality of God and the Universe'. Noble later describes Bach's 'Clavierubung' as 'the most perfect examples of the chosen genres that Bach, or arguably any other composer, ever produced. Noble is working within the Western, European 'classical' tradition, but it is the choice of the word 'perfect' that I find intriguing. It is used in a particular sense, which I think is similar to my definition earlier: 'in a condition which meets all perceived needs'. In this there are two important qualifiers. First, the music meets, or satisfies or fulfils. It does not need to do more than be sufficient for the task, although we would argue that at times it can seem to do so. Secondly, it meets all perceived needs. It does not look for needs beyond what are perceived by those involved with it, those listening to it. It does not need to predict future needs or remind us of previous needs; again, it is of here and now. Us, in this space, at the moment.

Placing myself within the experience and having reflected on it, almost as an example of action research, I have begun to develop my own notions of what happened, and their heuristics, and here is a brief summary:

- Both the Maasai and I accepted that there was potential for me to engage with the group.
- However, that engagement had to be negotiated, with me being prepared to be considered the novice and then initiand. This is in contrast to the tourist experience of being dragged from our seats to join in, whatever our preferences.
- There was a requirement for the development and extension of skills and knowledge, in songs, singing techniques, the ability to improvise and the preparedness to join in with movement. Part of this drew on, or was compatible with, existing skills; part was new.
- It took time, both immediately in terms of staying with the group for the period of singing songs (often several hours) and long-term in maintaining skills and showing progress over time, including time spent away from the group.
- It was time-specific, in being dependent upon *types* of time when performance was appropriate, if not the specific time, although type may vary significantly.
- It was site-specific, in being dependent upon *types* of site for performance, if not a specific site, although type may vary significantly.
- It involved the whole - body, intellect, mind etc.
- It was experience, thought and embodiment, which was aware, transformational and motivational.
- It was of significance to the participants.
- It was dynamic and susceptible to change at micro/macro levels.
- There was the possibility to engage at a preliminary level, while needing to be aware of further potential, which would need to be addressed at some point.
- This potential included the perfectable.
- Engagement happened at individual level, community level and both simultaneously.
- It led beyond *amour proper* (a defensive approach to the self) to amour de soi (a reasonable and open love of the self, after Rousseau [see Shaver 1989] and thence to 'Wi', indeed to an 'Adorable Wi'.

Of course, in finding this with the Maasai, I have become aware that this was also part of my earlier peak, spiritual or congregational experiences. I have found friends who take the same approach to performance as I do. There is enjoyment, in both process and product, but it is found through a degree of struggle, perseverance, challenge, openness, seriousness, provocation and so on.

CONCLUSION

The coding and decoding of significance, and its articulation, expression and interpretation, are not straightforward, even in such apparently small and self-contained communities. The practices described above are taken principally from rural communities and are still common there. This weighting is partly because a deliberate decision has been made to ensure this. My Maasai colleagues who have gone to university or to work (often in the armed forces) plan their lives so that work is done in one long period, with leave saved to allow as much uninterrupted time as possible in the home community. This allows a return, intellectual and spiritual, to identifiably Maasai practices that are felt to be nurturing and important to retain. Nevertheless, there are places where this has virtually been abandoned as the Maasai do not live in isolation. They are increasingly exposed to the same demands and policies as the rest of the populations of Kenya and Tanzania.

As people struggle to make sense of identities that are simultaneously individual, local, national and global, it may that the Maasai model - which locates cultural action as events, within asynchronous contexts reified as constantly improvising musical contexts, which require individual cooperation to create the communal whole (what I have synthesized above as the notion of 'wi') will continue to nourish and perhaps gain wider regard. The Maasai know this. They know they are equal with God, and this works because as an enculturated community they are an embodiment of God, singing being into existence.

I am grateful to my students studying 'Spirituality of Performance' who have engaged with this work vigorously and argued this chapter into a better form than it would otherwise have been.

REFERENCES

Barthes, Roland (1977) *Image-Music-Text*, London: Fontana.

Begbie, Jeremy (1991) *Voicing Creation's Praise: Towards a Theology of the Arts*, London: T & T Clark.

Boyce-Tillman, June (2007) *Unconventional Wisdom*, London: Equinox.

Butt, John (1997) 'Bach's Metaphysics of Music' in John Butt (ed.) *The Cambridge Companion to Bach*, Cambridge: Cambridge University Press, pp. 46-59.

Donovan, Vincent J. (1982) *Christianity Rediscovered: An Epistle from the Maasai*, London: SCM Press.

Floyd, Malcolm (1999) 'Warrior Composers: Maasai Boys and Men' in Malcolm Floyd (ed.) *Composing the Music of Africa*, Aldershot: Ashgate, pp. 135-64.

Floyd, Malcolm (2000a) 'Maasai Musics, Rituals and Identities', in Karen Ralls-MacLeod and Graham Harvey (eds) *Indigenous Religious Musics*, Aldershot: Ashgate, pp. 84-101.

Floyd, Malcolm (2000b) *Music in Enculturation and Education: A Maasai Case Study*, unpublished PhD thesis, Birmingham Conservatoire, University of Central England.

King, Anna (1996) 'Spirituality: Transformation and Metamorphosis' in *Religion* 26.4: pp 343-51.

Kipury, Naomi (1983) *Oral Literature of the Maasai*, Nairobi: Heinemann Educational Books.

Noble, Anthony (2008) *'Choirs of Larks and Tibetan Trumpets': In Search of the Holy Spirit in Music*, paper to be read at the International Baptist Theological Seminary, Prague.

Ralls-MacLeod, Karen and Harvey, Graham (2000) *Indigenous Religious Musics*, Aldershot: Ashgate.

Saibull, Solomon ole and Carr, Rachel (1981) *Herd and Spear: The Maasai of East Africa*, London: Collins and Harvill Press.

Shaver, R (1989) 'Rousseau and Recognition', *Social Theory and Practice* 15.3: 261-83.

Spencer, Paul (1965) *The Samburu*, London: Routledge and Kegan Paul.

Spencer, Paul (1988) *The Maasai of Matapato*, Bloomington: Indiana University Press

for the International Africa Institute.

Vaughn, W. (1994) *Romanticism and Art*, London: Thames and Hudson.

Local Diasporas / Global Trajectories
New aspects of religious 'performance' in British Tamil Hindu practice

ANN DAVID

INTRODUCTION

The tension and intersection of global and local influences play a significant part in the complex construction of identity and place-making within diaspora groups. Ann Cvetkovich and Douglas Kellner (1997:12) remind us that 'tradition, religion and nationalism' remain contemporary forces in the construction of both personal and national life, and it is these forces of tradition, religion and nationalism that have considerable impact on dance and movement practices and ethnic identity in British Hindu communities.

This paper discusses, through an ethnographic examination[1] of cultural and religious performance within British Tamil Hindu religious practice, the increase of performed religiosity in the diaspora setting. These performances appear to confirm and display not only a general Hindu identity but a specific Tamil religious identity, located, as they are, within Tamil temple ritual and at Tamil-specific festivals such as *Tai Pusam*[2]. Dance practices, including the classical style of Bharatanatyam and trance dance with body piercing are increasingly on show at such festival occasions. Additionally, the confident growth of new temples and their adaptation to British Hindu worship indicate an increase in local diaspora settlement and reveal how global trajectories are an essential factor in this expansion. The building of a new Tamil temple in east London, and the planning of a second one in the same area, is considered from this point of view. Are these new spaces for worship contested in any way? How has the space been appropriated? What is the significance of these newly-created Hindu religious places for the 'performance' of devotion? This paper seeks to question the adaptive strategies for preservation, modification and performance seen in British Hindu communities[3] that are fuelled by internal pressures within the communities and by outside forces, that is, by both local and global forces.

SRI LANKAN TAMILS IN EAST HAM, EAST LONDON

A focus on east London in the borough of Newham, where a large number of displaced Sri Lankan Tamils are settled,[4] reveals evidence of this new place-making centred on the Hindu temples that draws upon dance and movement practices alongside a growth in religious ritual and prominent outward display at religious festivals. We can see how place is transformed through religious practice and performance, indicating, as Bente Nikolaisen in her work on Mevlevi dervishes has argued, that 'place is as much about doing as it is about being' (2004: 98). Kim Knott echoes this point in her spatial analysis of the location of religion when she remarks that '[r]itual practice itself is interesting when seen from the perspective of spatial practice, as it is none other than spatial practice transformed by religious meaning' (2005: 43).

East Ham High Street is dominated by a Tamil presence that includes restaurants, retail shops of every description, solicitors' offices, money-lending facilities and a Tamil community

1 Initial fieldwork took place in Tamil temples and Tamil Saturday Schools in areas of Greater London, funded by the AHRC (Arts and Humanities Research Council).

2 This is an annual Tamil Hindu festival dedicated to Murugan, son of the deity Siva.

3 I am aware of the contested nature of the term 'community' and make use of the term consciously, while fully acknowledging the cultural, historic, geographical and religious diversity of British Hindu groups.

4 The London's borough of Newham in the east of the city is one of the most ethnically diverse boroughs hosting the highest proportion of non-white ethnic groups and the second highest groups of Asians in England and Wales (2001 census). 24.3 per cent of Newham's residents are Muslim, and 6.9 per cent Hindu. The Tamil population in the borough is estimated to be around 5-6,000. An even larger community of Tamils is resident in the London borough of Brent, located to the north-west of the city, with estimated numbers at around 12,000. Brent is one of London's most culturally diverse boroughs, where the

Performance Research 13(3), pp.89-99 © Taylor & Francis Ltd 2008
DOI: 10.1080/13528160902819364

• **Tamil shops on East Ham High Street, London.** *Photo Ann David*

non-white ethnic groups in the borough now form the majority of the population at 57 per cent; of the total, there is a Hindu population of 17 per cent (see Brent Council's website <www. brent.gov.uk> for further information). Tamils began to settle in the Borough of Brent in Willesden, Harlesden and Neasden in the 1970s, as many were students at Willesden College, and then began to move out to the areas of Kingsbury, Queensbury and Wembley. As the community grew more established, Brent became an attractive place for many more refugees, and now houses a vibrant Tamil community.

[5] The Tamil population is predominately Sri Lankan, although Singapore Tamils were the first to come to the area [personal communication with Newham councillor, Paul Sathianesan].

[6] Kaveri Harriss, in her breakdown of the largely working-class Muslim population of Newham (2006: 3), notes that the majority are Sunni Mulims, describing themselves broadly as Bareilvi or Deobandi. 85 per cent are of South Asian origin, and the remaining 15 per cent are from Africa, Europe and Britain.

[7] Worship, or worshipper of the Hindu deity Siva.

centre[5], despite the fact that Newham has a large Muslim majority.[6] Interspersed between these Tamil businesses are two mosques, a South Indian Hindu temple and various Middle Eastern banks in addition to other retail businesses supporting the wide ethnic mix that makes up Newham's population. Behind the main street is a Sikh Gurdwara, a newly built, traditional Saivite[7] Hindu temple, Tamil Catholic churches and other further evidence of a wide spectrum of religious worship. It is the strength of the transnational connections of these more-settled Tamil migrants that draws even greater numbers of Tamils to the borough seeking asylum, searching for employment and for places of refuge, and creating the second largest refugee community in the London Borough of Newham. Out of nearly 5,000 asylum seekers arriving in 2002 in Britain, the majority was from Sri Lanka, and it is estimated that nearly one quarter of the British Tamil population are asylum seekers. As far as religious practice is concerned, Tamils reveal the highest numbers of participation at Tamil temples or Tamil churches, and the support provided by their community in socio-cultural, religious and economic ways impacts directly on the settlement experience. Studies carried out reveal that virtually all Tamil respondents gave their reason for coming to Newham as kinship or friendship ties, or because of other refugees there and the presence of places of worship (Bloch 2002: 79).

Yet establishing these local places of worship has not been without controversy. One Hindu temple on the main High Street - the London Sri Mahalakshmi Temple, founded in 1985 and set up in the premises of an old clothing shop - transformed its interior into a small, traditional south Indian temple with 'faux bas-relief' pillars fixed onto the outside walls. The managing committee of this temple has purchased a disused Public House 200 yards away on the High Street at one of the main road junctions and has submitted plans to demolish the building and erect a three-story traditional Hindu temple on the site. The land and property cost £1.7 million (in 2004) and charges for the new and ambitious building are estimated to be around £6 million. The new basement will house kitchens and a dining area for up to 400 people; the temple will be located on the ground floor and will accommodate three major and eight smaller shrines, all constructed according to traditional practice; and the top two floors will provide cultural, social and medical centres for the devotees, including an auditorium for music and dance performances (with stage and seating).

• **The new London Shri Murugan Temple, East Ham.** *Photo Ann David*

This will be the second traditional purpose-built south Indian temple in London, and the third overall in Britain.[8] Local residents, however, at a specially held meeting with Newham Council, voiced the usual concerns over parking, about increased congestion in the area, particularly at festival times, and questioned why another Hindu temple was necessary in an area with many already existing religious centres. Planning permission was finally granted by the local council after certain conditions were agreed,[9] and a second, revised, application had been submitted.

Concerns too have been raised by local residents about the site of a Sri Lankan-run Tamil Hindu temple in Wimbledon, southwest London. Their focus is the use of space and an encroachment into what is considered a secular, public and suburban city area. The temple is not seen by these non-Hindus as an asset to this mainly white, urban place whose residents remain uneasy but now tolerant of the outdoor festivals processing through the local streets, as discussed later. Anthropologist John Eade (1996: 225) points out that disputes and criticisms in contested spaces are often 'motivated by cultural exclusion' and draw on 'issues of what is culturally acceptable'. In Shi'a Muslim groups in Ontario, research studies revealed the problematic nature of the creation of sacred space by religious groups. By constructing space that is exclusively sacred, the risk is that the world outside of that space is then made exclusively secular (Schubel 1996: 202). These local concerns situate themselves within a growing sense of global religious identity, where concerns exist regarding the separating of culture and religion from local sites of practice and settlement.

TRANSNATIONAL CONNECTIONS / GLOBAL FLOWS

The strength of the geographical movement and transnational ties between members of the British Asian diaspora creates a flow of people, capital, material goods and ideas. These global flows reveal how intersections between the local and global affect cultural and religious practices, exemplified by the case of customs from folk Hinduism, such as body piercing and trance dancing, being exported throughout the Tamil diaspora. Such rural traditions are incorporated into new local practices in east London, as I discuss later. There is a regular two-way circulation between Europe and Sri Lanka and India by members of Tamil groups here, despite the continuing violence of the civil war in Sri Lanka. Current studies being carried out by the author at a newly established Tamil Hindu 'worshipping centre' in East Ham reveal that many of the attendees are relatively recent Tamil migrants from other parts of Europe - Germany, Holland, France and Switzerland - and are frequent travellers between Britain and Europe and between Britain and Sri Lanka and India.[10] The move into the new millennium showed a change in Sri Lankan Tamil migration patterns with a marked decline in asylum migration but an increase in the regrouping and relocating of family groups, particularly of Tamils from Europe to Britain. This secondary migration has perhaps been pushed forward by desires for the younger generations to gain an English education and by the obvious support of a well-established, significant British-based Tamil community.

The Sri Lankans' political awareness and cultural ties to their homeland are major factors in their diasporic situation, in the sense of a displaced people whose political and financial support is devoted to rebuilding and supporting their homeland and whose sense of a global, diasporic family is a powerful articulation of Tamil Sri Lankan nationalism (David 2005). It is interesting to consider here the concept of cultural and religious place-making for the Tamils in terms of Tamil Eelam. Not only is there evidence of place-making in the diaspora, but for many Tamils the establishment of their own state in Sri Lanka absorbs much of their emotional, political and financial capital. As with many diaspora groups, attention is turned two ways - to the past and to the future, in two contrasting places, and to the local and the global.

8 Interview with the temple president Dr P. Alagrajah, at this temple. The two other temples are the London Sri Murugan Temple in East Ham and the Shri Venkateswara (Balaji) Temple of UK, both newly built according to south Indian architectural traditions.

9 The conditions were that no weddings would take place in the temple; prior approval must be obtained for kitchen ventilation and extraction; prior approval of materials and of an environmental code; and that the main entrance would be sited away from the major traffic lights at the road junction (see notes of Newham Council meeting and public forum meeting July 2005 <www.newham.gov.uk>, accessed 12 March 2007).

10 This is the Melmaruvathur Weekly Worshipping Centre in East Ham, east London and is part of an ongoing study examining the religious lives of migrant groups. See <www.surrey.ac.uk/Arts/CRONEM> and note 11 below.

DANCE AND MOVEMENT PRACTICES

If we examine this cultural flow in terms of dance practices, in recent years the accessibility and lower costs of international travel have enabled the dancers and dance teachers I interviewed to travel frequently to India and Sri Lanka to purchase new dance costumes, to record fresh music tapes or to have extra lessons with their original dance *gurus*. Thus such transnational travel informs, supports and makes possible the local manifestations, in this instance, of South Asian dance performance and dance transmission in Britain.

In the Tamil community in East London, the scene appears at first glance to be more localized. The transmission of Bharatanatyam, a now-global dance style, remains dominated by local modalities in the predominately Sri Lankan Tamil temples and Tamil weekend schools. Students are neither encouraged to attend international performances of South Asian dance nor are made aware of the work of well-known performers in London and the rest of Britain. Most syllabi and dance classes are taught in Tamil, even though the second-generation students are more familiar and more at ease with English. Despite the influences of London as a major global city, it is these local factors that influence the Tamil community. The more inward-looking and somewhat wary nature of the Sri Lankan Tamils and their more recent immigration and settlement patterns are factors at work here.

The global influences of Bollywood dance styles (see David 2007), for example, sit uneasily with the first-generation's adherence to their traditional cultural and religious beliefs, and there remains a tension between the struggle to maintain local (and Tamil) cultural traditions against these perceived eroding forces of globalization. This 'production of locality' (Appadurai 1996: 180), seen in these terms, demands constant work to be sustained under the pressures of diasporic living. Appadurai notes how, in terms of local cultural reproduction, 'space and time are themselves socialized and localized through complex and deliberate practices of performance, representation and action' (180), thus creating names, properties, values and meanings for the local community through their specific cultural symbols.

We see here an ongoing persistence of local cultures in the face of powerful globalized forces. These local practices perhaps at times sit uneasily alongside global culture, maintaining a 'living' place and space in community life yet at other times appearing to be an essential part of both national and global trajectories. The notions of national and local, as both a cultural and theoretical construct, are at once binary and singular, embracing the local within the national but also creating contested and political pathways. This intersection of local issues with national and transnational processes stretches across the class divide with both the well-off and, as Eade argues, 'the comparatively poor (materially) labour migrants in London [being] highly adept at negotiating landscapes and constructing imagined worlds which cross local and national boundaries' (2002: 129).

PERFORMING THE DIVINE

Exploring concepts of place and space, their production and their dynamics in relation to the body and its social relationships, has drawn academics from a wide range of disciplines such as social anthropology, geography, religion, performance studies and dance. The centrality of the body in space in social, cultural, political, economic and religious life impacts onto scholarly understanding across these areas, creating questions of the representation of space, the use of space, the individual's experience of space and the types of discourse at play within it. We are dealing here with not only physical space but the emotional, social and subtle concepts of spatial existence. Henri Lefebvre reminds us that 'it is by means of the body that space is perceived, lived – and produced' (1991 [1974]: 162). For many communities, sacred and secular space are significant notions and ones that are often

• **Hindu priests at the London Shri Murugan Temple, East Ham.** *Photo Ann David*

bounded by dance and movement practices, as well as by the interplay and exchange between public and private space.

Yet how does the building of new temples within the British Hindu community conform to a sense of sacred space when they occupy, for the most part, crowded and contested urban sites, hemmed in by other surrounding buildings and dominated by housing rather than the elements of nature? Joanne Waghorne suggests this may be the reason for the 'increasingly elaborate consecration rituals now, which sacralize both the temple and its land' (2004: 235). The change in temple structure to accommodate the British climate, incorporating such items as under-floor heating, carpets and glass windows, and the changing patterns of British Hindu community worship (needing larger communal spaces) while retaining the sacred geometry of the temple within the traditional architectural formula, is a phenomenon still to be researched, and the question of the creation and use of sacred or ritual space remains a pertinent one.

Within the Hindu temple, the ritual space that exists in front of and around the deities is mediated by the male priests, and contact with the deities is solely through their agency.[11] Before entering the temple each morning for ritual activity, the priests undergo a process of transforming their bodies physically, mentally and spiritually, by bathing, reciting mantras and using specific hand gestures. Mantras are imposed onto the hands, constructing a new, divine state. In this way, the priests are purified and their bodies are perceived then as sacred space. Different spaces in the body are specifically and spatially located, not only in their physical presence but indicated too in a metaphysical mode. This is a complex layering and transition of an individual and physical body into what is conceived as a pure receptacle and vehicle for divine powers.

Hand gestures with symbolic and cosmic significance are employed not only in this bodily transformation but additionally in temple ritual by the priests, and many of these gestures appear too in the dance system. Lefebvre writes of how such 'gestural systems connect representations of space with representational spaces' (1991 [1974]: 215), noting how the systems signify worlds codified by social and religious practice and how they create their own socio-cultural and religious space. Kim Knott, examining the location of religion through use of a spatial analysis, states that it is through the body and the hands in particular that 'cultural acts of separation and signification can be achieved' (2005: 223). Body and territory are formative for the notion of the 'sacred', and sacralization produces distinctive, and contested, spaces. Knott's discussion of Lefebvre notes that 'the body, through a spiritual gestural system, has produced a new space – one that is physical, social and symbolic' (2005: 135).

[11] Current research by the author examines the unorthodox practice of women conducting the temple rituals and acting as mediators between the deities and the devotees. One example is the Melmaruvathur Amman Temple, based in India but now with branches all over the world, which actively promotes women to take on key roles in ritual. A group in East Ham, set up in 2001, meets in a converted house three times a week for worship, and although men attend, the ritual is run by women and it attracts mainly women followers. From fifty to sixty attend each meeting and on special occasions 2-300 devotees may be present.

• **Temple chariot with deity in the local streets of East Ham.** *Photo Paul Sathianesan*

Space is ordered through 'the way people move within the space they occupy' (Ebin 1996: 103), creating what I have called 'a choreography of the temple'. Hindu devotees will move in the sacred space of a temple in a practised, transcribed mode, circumambulating clockwise around the specific deities, prostrating the body on the ground[12] before the principal shrine, making particular gestures of the hands and successive bending of the knees before other gods. Not only are the bodies in this space performing an unconscious choreography, but an inner rhythm is created, focusing the attention and transforming secular time and space into a ritual, sacralized locale. Deirdre Sklar's investigation of the Catholic fiesta in Tortugas, New Mexico, discovered too that by examining the detailed actions and 'doings' of the participants 'clues to faith, belief and community would be embedded in the postures and gestures of the fiesta' (2001: 4). She notes that by joining the community in the preparatory actions for the fiesta, whether chopping vegetables for the feast or dusting the large statue of the Virgin Mary, she 'entered into communal sacred time, for that is what the Tortugas fiesta was' (183).

When the temple deities are brought out into their urban, London and often white, middle-class surroundings from the temple for festival processions, such as the annual chariot festivals *(ter)*, these streets too are transformed into sacred/ritual space. At these times, local people maintain what Martin Baumann has termed 'a distanced toleration' (2002: 93). The devotees walk barefoot, as shoes are taboo in a purified, ritual space, and some will even perform *angapradhakshanam* (ritual body-rolling) along the streets behind the chariot. One man I witnessed undertook a vow to roll prostrate on the ground around the four local streets of the Shree Ghanapathy Temple, in Wimbledon, southwest London. He wore a cotton *dhoti*, tied around the waist and above the knees, with the rest of his body bare, and no extra protection for his body. The procession lasted for at least two hours, as the chariot made slow progress, stopping every few yards for the devotees to perform *puja* to the deity. These actions, together with the presence of the deity in the chariot, create a new, temporary spatial sacredness in secular, populated urban space. Within the temple itself, the sacred yet public space is animated by the ritual movement of the priests and the worshippers and is enhanced by the powerful sensory nature of Hindu worship - the rich scents of incense hanging heavy in the air; the ringing of bells and the sounds of ritual chanting and temple instruments[13] being played; the stunningly bright colours of saris, dhotis, shawls, decorations and burning lamps transforming the space into one where ritual and sacred communication may take place.

When dance performances occur in the temple, pertinent questions arise relating to the dancer, the devotees and the deities in relation to sacred space. For whom is the performance? For the deities? Or for the devotees, or for both? Is it

[12] This is performing *astanga pranam*. Men will lie so that their full body is face down on the ground, allowing eight significant points in the body to touch the floor. A woman's body has five such points, so she will kneel and allow her arms and head to touch. This also keeps the woman's pelvis from the point of contact on the ground, as this part touching the ground would be considered inauspicious.

[13] It is common to hear temple musicians playing the traditional instruments of *nagaswaram*, a double-reeded flute (like an oboe), the *tavil*, a large, outdoor drum beaten with a curved stick one side and the drummer's hand with metal covers on the fingers on the other, and the Indian cymbals (*talam*) at festival occasions. They play as an accompaniment to the deities, at times of ritual worship in the temple and at festival times, to initiate processions, and as a prelude to the deities' arrival on the streets during processions. As their sound is so powerful, they are considered to be outdoor instruments.

acceptable for the dancer to have his or her back to the deities?[14] Several dancers have spoken to me of the potency of the moment of contact with the deities by facing them and dedicating the dance to them when performing in the temple's sacred space. At British Bharatanatyam performances for *Navratri* (a nine-night autumn festival celebrating the feminine power of the divine) in the Mahalakshmi Temple in East Ham, east London, the dancers performed in the small ritual space directly in front of the main deity. They performed with their backs to the deity and therefore facing the devotees, a characteristic perhaps of staged auditorium performance. A professional Odissi dancer who performs in temples both in Britain and in India described dancing so that her back was never towards the deities - not always easy to achieve when there is limited space. She commented too on performing at the newly-established temple dance festivals in India where performances take place outside of the temple on specially created stages. The concerts are held in the evening and are attended by many hundreds of people, seated as if in an outside auditorium. The dancers, if they wish, may perform their own dance *puja* in front of the deities before going on stage, in the inner hall of the temple rather than the sacred inner sanctum, and away from the audience.

In the same way, classical Indian dancers perform an invocatory bow on stage or in the wings before a recital begins, in order to transform the stage into a sacred space. In her report on the project *South Asian Dance in Britain: Negotiating Identity through Dance* (2002),[15] dance scholar Andrée Grau noted that by

> observing the way dancers approach the dance space shows another kind of link with spirituality. In many genres space is consecrated by the dancer's invocation, for example, and this invocation transforms it so that it can be inhabited in a special way. (Grau 2002: 70)

Dancer and choreographer Akram Khan is aware of an inner tension when performing his contemporary dance work, as the 'feeling' of the stage as a temple, as sacred space, which he experiences when dancing the classical Kathak style, is not present. He questioned how he could draw the two together to diffuse the tension:

> For me, when I enter the stage, it becomes like a temple. It's just natural to me because of the tradition and training in classical dance ... Everything on stage is always living. It's been made by a human being and is danced by a human being, so there is an inherent spirituality there anyway. The question is how to convey that in a more 'abstract' piece ... Indian dance has spirituality embedded in it. (Khan, interview with author)

These comments suggest how by sacralizing the space for performance through a creative ritual process, a dancer makes a meaningful place which can then be inhabited by the dancing

[14] If we look to examples in the Roman Catholic Church, the original performance of the Mass privileged the priest and his relationship with God, rather than the congregation. The priest recited the Mass in Latin, in a low murmur hardly heard by the attendees, and faced the altar with his back to the congregation. The priest was communing with God, rather than with the audience of worshippers, and remained in the reserved sanctuary space (the sacred space around the altar). This changed radically in 1963 with the Second Vatican Directive, which brought a new theology of the Mass, offering a democratization of the procedures of worship. From then the priest faced his audience, spoke to them in English, and was allowed to move out of the sanctuary to give communion. [Personal communication from Brendan McCarthy].

[15] This two-year research project ran from 1999-2001 and was funded by the Leverhulme Trust. It was headed by Dr Andrée Grau and carried out under the auspices of the Centre for Dance Research at Roehampton University.

• **Bharatanatyam dance performance at Navratri festival, East Ham.** *Photo Ann David*

body, trained as it is in a system that privileges 'spirituality' or a faith in God. As Knott remarks, 'Body and territory are formative for conceiving of the "sacred"' (2005: 103)

PERFORMANCE AS TRANCE

Further examples of transformation of the embodied form of the dancer are shown in the increasing display of trance-dancing and bodily mortification such as piercing seen at public Tamil Hindu festivals in London. Onlookers and participants describe this as a state of possession, and a display of the power of the deity. The unpredictable nature of possession in that it can happen during *puja* (worship) or while singing *bhajans* (devotional songs) or in receiving *darshan* from the deity allows it to be accepted and honoured in Hindu society, demonstrating that it is possible for a devotee to become united, albeit briefly, with his favoured deity. This is the sought-after intimacy between worshipper and deity which is the common goal of Hindu worship.

Dancing while possessed is common during some of the important Tamil festivals, particularly the annual August chariot festival, as well as the January *Tai Pusam Kavadi* festival. Men devotees undertaking a *kavadi* vow *(vrata)*[16] are in a state of possession, where not only do they dance, but they are believed to be possessed by the deity who will in turn be pleased by their state of purity and devotion and will grant them their boons. Vows are made by the men for their families or friends, to dispel sickness and troubles or, more altruistically, for the sake of the world. Some may choose to undergo body piercing. Before beginning the body piercing, the dancing, and the carrying of the heavy weight of the wooden *kavadi* on their shoulders, the men will have undertaken various purifying rituals for the days leading up to the event. Fasting, bathing and praying will have been rigorously practised to enable a condition for the deity to take possession. Then on the day itself, this state of purity and the devotion practised empowers the men in such a way that they are said to feel no pain when their skin is pierced. In fact, Kapadia notes that in matters of possession

> [r]itual purity is of secondary importance: it is merely necessary so that the person might be a fit vehicle for the deity, and so certain simple measures (involving vegetarianism, chastity and extreme cleanliness) are taken by the person seeking to be possessed. But it is devotion that counts above all. (2000: 183)

The benign ecstatic trance dancing of the Tamil devotees at the *Tai Pusam* festival is a potent signifier of Tamil devotion, a performance of faith in the deity Murugan, expressed in the movements and gestures of their dance. It reveals clues about the nature of faith in Tamil Hinduism, that is, in this instance, a belief that the powers of the deity Murugan become embodied in the purified body of the devotee during the dance. These same powers are also believed to bring auspicious results for a vow or plea for help made by the dancing devotee - factors that are specific to that community. It becomes, too, as does the dancing form of Bharatanatyam, both a local and national identity marker, appearing in a globalized and localized diasporic location - local in its adherence to Tamil Hinduism and even more so, to Sri Lankan Tamil Hinduism, and nationalistic in its adherence to Hinduism in general.

It is a well-recognized fact in the Tamil Hindu community that piercing, self-mutilation or the ritual practice of fire-walking are proof of the existence of a 'possessing deity whose protective

16 Vrata is a voluntary religious vow, which forms an important part of popular Hinduism.

• Carrying Kavadi at the Tamil Tai Pusam festival, East Ham. *Photo Ann David*

powers prevent any harm to the dancers' (Blackburn 1981: 217), although men and women are watchfully present in case of any human error. Stuart Blackburn writes of a localized oral tradition in Tamil south India called the 'bow song', which takes its name from the 6- to 12- foot-long bow instrument that is played, along with several other instruments. Performances take place at festival times and are accompanied by actors who go into a state of possession during key points in the narrative. The possession dance is the 'central focus of the entire festival' and is known as 'god dance', and the medium is described as the 'god dancer' (216). Blackburn adds this description:

> The possession dance is also a register of the affective intensity or 'ritual depth' of the festival. Ritual depth refers to the relative level of emotional power and ceremonial import that underlies the event. Although essentially a subjective quality, it is unmistakable precisely at those points of the narrative and ritual convergence. There it exhibits certain observable features in its primary expression as dance; significantly, although people can identify possession by its inner state ('frenzy'), they always refer to it simply as 'dance' (attam).
> (Blackburn 1981: 217)

Andrew Willford's research on the Malaysian Tamil community found that 'a resurgence of Tamil Hindu consciousness is being expressed through dramatic rituals which often include practices of self-mortification' (Willford 2002: 247) and that through the festivals such as *Tai Pusam*, a greater 'Tamilness' is being signified and asserted in this Islamic state. Willford emphasizes how the main avenue for expression of Tamil ethnic identity is through cultural and religious practices. This is especially the case as the Tamils are concerned about their perceived loss of cultural, economic and political rights. An 'increase in the size and intensity of a number of festivals, a proliferation of urban shrines and a growing clientele for spirit mediums' is evident, notes Willford (255). Bearing the *kavadi*[17] for Lord Murugan has grown rapidly in importance in Malaysia, and the 1995-6 *Tai Pusam* festivals attracted around one million people. Large numbers of men and women will go into a trance and perform energetic dances, the men enacting 'a feverish traditional circular-step *kavadi* dance' (260).[18] My recent visit to the site (November 2007) confirmed its status as a major pilgrimage site for both Hindus and on a lesser scale, non-Hindus.

The above examples differ in significance and purpose from the phenomenon of trance dance in other parts of the world. Bruce Kapferer's work among Buddhists in Sri Lanka (1991) analyses some of their powerful exorcism rituals where trance dancing is the specialism of the team of male exorcists. Accompanied by drummers, chanters and assistants during the all-night ceremony, the men's bodies become possessed by the demonic forces inhabiting the 'patient' to be exorcised, and an elaborate, competitive and fierce dance proceeds. The 'patient' himself may or may not enter into a semi-trance. Balinese dance too contains many instances of trance or embodied dance, particularly as part of the sacred forms performed in the jeroan, the inner temple area. Embodiment is commonly by divine or occasionally demonic spirits and the practice varies according to locality and to the particular type of *Sang Hyang* dances. A common form is danced by prepubescent girls, whose bodies are embodied by divine nymphs once the girls are in a trance. These dances are performed at irregular intervals when needed, to ward off danger or to mitigate disasters.[19]

Theorists of states of trance and possession (Rouget 1985, Bastin 2002) speak of the need to distinguish analytically between trance (a state of altered consciousness) and possession (where the human state is temporarily displaced or inhabited), although, as I have found in my own examples and the illustrations above, both can be experienced together forming a single, composite, embodied phenomenon. Embodiment by a spirit or by a form of a deity constitutes a specific performance in which the body of the individual literally becomes the body of the possessor - 'embodiment' being a more accurate

17 The name of the wooden structure male devotees carry on their shoulders during festivities.

18 Willford notes that more modern tunes, which have originated from Tamil films, are now being played for the dancing and have become accepted as devotional songs. He adds that '[t]o some extent, the *Kavadi* dance itself has evolved to now include steps and gyrations from the popular India-produced films' (2002: 260).

19 The Balinese language does not have simply one word for the widespread phenomenon known as 'trance'. Several terms are used: *karauhan kalinggihan, kalinggaan, kodal* or *tedun*. These may be translated in different ways, such as 'a (temporary) loss of the soul' or 'a state with another spirit other than your own' - rauh being the Balinese word for spirit.

and more widely used term than 'possession'. This state of embodiment is indicative of socio-cultural beliefs that identify powerful forces and influences outside the individual, in contrast to a Euro-American culture that identifies them within the individual. Possession involves present experience, but one that is mediated by a historical mythology, and provides a form where personal and the collective are yoked together, as the individual internalizes the form of the deity. Public witnessing is of great significance to the event, allowing a corroboration of the extent of the possession. Although embodiment and trance dances contain multi-layered significance that varies from group to group, for Hindus the prime attribute is one of purity and auspiciousness. The possession is a statement, an exhibition of the 'moral status' (Kapadia 2000: 181) and level of purity of the caste group or community and hence its importance. It is common practice for lower-caste Hindu groups to seek to raise their hierarchical status by adopting customs of greater purification such as fasting, ritual bathing, donating of gifts, becoming of a higher status. The system is one of upward mobility, based on concepts of purity, and hence the significance of an event demonstrating greater purity or auspiciousness.

CONCLUDING REMARKS

The new sacralization of public and private space, both architecturally and in an embodied sense, continues to be part of British Hindu religious practice, and evidence appears to suggest, as I have argued, this is on a growing scale. Dance and ritual movement forms are playing an increasingly significant part in this changing use of sacred space and the growing religiosity of temple ritual, indicating a new trajectory of embodied performance of Hindu faith or religion.

As we have seen, a group's cultural, ethnic and religious identity may be articulated through dance, acting as a symbol of both continuity and change, and its presence in the diasporic setting can be a powerful marker of the group's distinctiveness and place-making strategies. The expressive South Asian cultural forms rely heavily on dance, drama and music for their enactment, both in a secular and in a religious milieu, and the performing of religion and entertainment has spawned a multitude of embodied styles and disciplines. The dancing body stems directly from a religious discourse, and the concepts of that heritage, in terms of sacredness, still remain. Dance and related movement styles form an essential part in the production or performance of faith within Hindu worship - practices that are perhaps accentuated in a diasporic setting where certain cultural patterns that may be at variance with the local community have to be consciously transmitted. Performing faith is a way of actualizing one's faith, of making it a reality in an embodied form.

Through religious practice, the use of the body and its gestures and movements play a powerful part in the territoralization of space, transforming the elements of power and politics in a diasporic community's negotiation with its environment and revealing influences that are both local and yet informed by global processes. Tamil Hindu groups in Britain are gaining confidence in exhibiting bodily practices and religious worship in more public and often contested spaces, indicating a great stability and sense of power in their relationships with their local environments. Perhaps they have negotiated a space that is now marked and made their own and through this are gaining a spatial authority where they no longer feel culturally displaced. An understanding of how people perform their faith in space and place and the contested issues surrounding it adds to the global debate and interest in South Asian religious practice and its cultural representation that is rapidly growing. Finally, it speaks of a past that is being transmuted in the present, evolving, growing and, I would argue, at an important moment of transition.

ACKNOWLEDGEMENTS

Initial research and fieldwork in Leicester and London were carried out with the support of a three-year grant from the Arts and Humanities Research Council (AHRC). Recent further research work has been funded by the Ford Foundation through the Social Science Research Council, USA (SSRC) and is part of a wider international and comparative research project on transnational religion. The author would also like to thank colleagues at the CORD (Congress for research on Dance) in New York, November 2007, for their comments on an earlier draft of part of this paper.

REFERENCES

Appadurai, A. (1996) *Modernity at Large: Cultural Dimensions of Globalization*, Minneapolis, University of Minnesota Press.

Bastin, R. (2002) *The Domain of Constant Excess: Plural Worship at the Munnesvaram Temples in Sri Lanka*, New York and Oxford: Berghahn Books.

Baumann, M. (2002). 'Migrant Settlement, Religion, and Phases of Diaspora - Exemplified by Hindu Traditions Stepping on European Shores', *Migration* 33(34/35): 93-117.

Blackburn, S. H. (1981) 'Oral Performance: Narrative and Ritual in a Tamil Tradition', *Journal of American Folklore* 94(372): 207-27.

Bloch, A. (2002) *The Migration and Settlement of Refugees in Britain*, Basingstoke, Palgrave Macmillan.

Cvetkovich, A. and Kellner D., eds (1997) *Articulating the Global and the Local: Globalization and Cultural Studies*, Oxford, Westview Press.

David, A. R. (2005) *Performing Faith: Dance, Identity and Religion in Hindu Communities in Leicester and London* [Unpublished Ph.D. thesis], Leicester: DeMontfort University.

David, A. R. (2007) 'Beyond the Silver Screen: Bollywood and Filmi Dance in the UK', *South Asia Research* 27(1): 5-24.

Eade, J. (1996) 'Nationalism, Community and the Islamization of Space in London', in B. D. Metcalf (ed.) *Making Muslim Space in North America and Europe*, Berkeley and Los Angeles, University of California Press, pp. 217-33.

Eade, J. (2002). 'Adventure Tourists and Locals in a Global City: Resisting Tourist Performances in London's East End' in S. Coleman and M. Crang (eds) *Tourism: Between Place and Performance*, New York and Oxford, Berghahn Books, pp. 128-39.

Ebin, V. (1996) 'Making Room versus Creating Space: The Construction of Spatial Categories by Itinerant Mouride Traders', ed. B. D. Metcalf, pp. 92-109.

Grau, A. (2002) *South Asian Dance in Britain: Negotiating Cultural Identity through Dance (SADiB)*, Roehampton: University of Surrey.

Harriss, K. (2006) *Muslims in the London Borough of Newham* [Background paper], Oxford, Centre on Migration, Policy and Society: 1-33.

Kapadia, K. (2000) 'Pierced by Love: Tamil Possession, Gender and Caste', in J. Leslie and L. McGee (eds) *Invented Identities: The Interplay of Gender, Religion and Politics*, New Delhi, Oxford University Press, pp. 181-202.

Kapferer, B. (1991) *A Celebration of Demons: Exorcism and the Aesthetics of Healing in Sri Lanka*, Oxford: Berg.

Knott, K. (2005) *The Location of Religion: A Spatial Analysis*, London: Equinox Publishing.

Lefebvre, H. (1991 [1974]) *The Production of Space*, Oxford: Blackwell.

Metcalf, B. D., ed. (1996) *Making Muslim Space in North America and Europe*, Berkeley and Los Angeles: University of California Press.

Nikolaisen, B. (2004) 'Embedded Notion: Sacred Travel among Mevlevi dervishes' in S. Coleman and J. Eade (eds) *Reframing Pilgrimage: Cultures in Motion*, London and New York: Routledge, pp. 91-104.

Rouget, G. (1985) *Music and Trance: A Theory of the Relations between Music and Possession*, Chicago and London: University of Chicago Press.

Schubel, V. J. (1996). 'Karbala as Sacred Space among North American Shi'a: "Every Day Is Ashura, Everywhere Is Karbala"', in B. D. Metcalf, (ed.) *Making Muslim Space in North America and Europe*, Berkley and London: University of California Press, pp. 186-203.

Sklar, D. (2001) *Dancing with the Virgin: Body and Faith in the Fiesta of Tortugas, New Mexico*, Berkeley: University of California Press.

Waghorne, J. P. (2004) *Diaspora of the Gods: Modern Hindu Temples in an Urban Middle-Class World*, Oxford and New York: Oxford University Press.

Willford, A. (2002) 'Weapons of the Meek': Ecstatic Ritualism and Strategic Ecumenism among Tamil Hindus in Malaysia', *Identities: Global Studies in Culture and Power* 9: 247-80.

The Rhetoric of Ritual:
Transformation as revelation and congregational liturgical dance as performance theory

CLAIRE MARIA CHAMBERS BLACKSTOCK

While the recent history of Christian liturgical dance in the United States is fairly brief (liturgical dance received much critical attention in 1963 after the publication of the *Constitution on the Sacred Liturgy* by the Vatican), human impulses linking dance to the sacred are timeless. Looking to the dancing congregation at St. Gregory of Nyssa Episcopal Church in San Francisco, California, this paper seeks to understand liturgical congregational dance as theory-making practice danced out in theory-making space. This church employs a 'rhetoric of ritual' that expands their 'theory' of divine community into a living, thriving reality. This liturgical 'work of the people' is also the work of theory. For the performance scholar, St. Gregory's explodes with potential for understanding performance afresh. Seemingly distinct concepts such as body/mind, divine/human and theory/practice begin to mesh. By dancing with this congregation, one listens to their story and participates in their struggle; one is claimed by it; it assumes one's response and one's responsibility. Christian or no, it becomes not a question of belief but of action and participation.

In many ways, St Gregory's is not an unusual congregation. Like most Episcopal churches in the United States, they - and I want to note here that I will use the collective in referring the congregants and the congregation - are predominantly Caucasian but celebrate their small diversities by flinging open wide, welcoming doors. They welcome the stranger and visitor with warm handshakes and building tours. Everyone wears a nametag. They serve their community with a variety of ministries, most notably their thriving food pantry. A visitor might notice the unconventional layout of the worship space, and the murals on the walls, but other than that, the place is pretty normal. It seems so ordinary that it takes a while for the *extraordinary* nature of their liturgy to sink in. St Gregory's speaks, sings and most importantly dances liturgy. This dance is fundamental to the central argument of this paper in which the dancing congregation, through a rhetoric of ritual, metonymically relates their liturgical practice to their worldly commitment.

The dance reveals that the work of the liturgy is not something that stops with the liturgy itself but continues in the outside community. In the same way that rhetoric in language is the effective and persuasive use of words, the rhetoric of ritual effectively and persuasively uses rhetorical elements such as dance, gesture, word, visual art, music, food and incense to demonstrate liturgy as a communal practice that moves beyond the repetition of a creed to an argument for a radical reality. To be fully present in the liturgy is to be fully present to the entirety of life, inside and outside the church building, and this is to practice the presence of God. The dance, therefore, is a metonym for the life of the church in its relationship to the larger world it serves, challenges, and in which it lives.

St Gregory's straddles a steep hill between residential and industrial areas. Its octagonal tower, topped with a Coptic cross and shingled in

Performance Research 13(3), pp.100-108 © Taylor & Francis Ltd 2008
DOI: 10.1080/13528160902819372

brown wood, is distinctive but blends in with the surrounding structures. The building is broad and welcoming, both round and square, earthy browns set off by red, blue and gold. Entering the front door before mass, any seasoned church-going visitor will recognize that this is a unique space. Immediately to the right of the entrance is a kitchen, from which people in aprons bustle and rich smells emanate. There is no narthex or entryway; the front doors deposit you right into the worship space, where the altar is the centerpiece in a large, airy, light-infused rotunda. Off of this circular space is a rectangular area, still tall and many-windowed, where a lectern

© David Sanger / www.davidsanger.com

stands on a low platform. At the other end is the priest's chair, and directly above this is a tall alcove that reaches up to the raftered ceiling. In this alcove is a mural of the heavenly bridegroom marrying the soul, while Mother Church / Mother God looks on and St Gregory stands below. A pair of glass-paned doors lead off the back of the rotunda, beyond which is an outdoor baptismal font, hewn from grey stone and planted with cool greenery. Incense mixes with the food smells coming from the kitchen. Sunlight glints on glass oil lanterns hanging above the altar; Byzantine-style icons on the walls, patterned rugs on the floors and ribbons hanging from processional crosses are all richly colored. Simply entering the worship space is a dance of sensorial perception.

A unique element of St. Gregory's worship space is that the walls surrounding the circular altar are covered in a vibrant mural depicting saints dancing around the perimeter of the room - hands on shoulders and knees raised, echoing the congregation's dance. These figures represent canonized saints but also people the community have chosen as their own 'saints', such as Martin Luther King and Martha Graham (only two of the eighty-eight figures in all). The dancing saints encircle the dancing congregation, framing the work of these people in a special way: the dance frames the frame of the liturgy, of the institutional church, but at the same time explodes the frames and the framing process by affirming that that which seems impossible - people and saints dancing together as one divine body both within and without the institutional church - is in fact a reality. We see that the framing of performance is inseparable from that performance itself. The framing process is the experience. In the ritual framing of the dance, the people express longing for what could and should be, but in that longing actually break through the frame to experience a new reality. Here, where faith and practice intersect, is the marriage of liturgy and performance theory.

The weekly liturgy in ordinary time begins with music around the altar (to remind us of the centrality of the table and Christ's hospitality, perhaps). Then we process, walking with be-streamered crosses, liturgical umbrellas and music-makers to the seating area, where the liturgy of the word begins. The liturgy includes vocal directions from the leaders, so no one is ever confused or lost. The readings may inspire silence or laughter, and the sermon is shared by the congregation, who take seriously the priest's invitation to preach for and to each other by offering personal reflection or anecdote. The transition from word to table is through a travelling dance: placing hands on shoulders of the person in front of us, we sing a hymn to a rocking step-dance until we've all encircled the altar. The liturgy of the table is all responsorial song. The congregants administer the bread and wine to one another, passing the bread and cup from hand to hand, saying each others' names with purpose: '*Claire*, receive the body of Christ.' After the Eucharistic feast we join together for the final dance, circling the altar with lifted knees in a modified grapevine step, children held aloft on shoulders, drums, rattles and the clear song of simple voices in unison vibrating the room.

This liturgy of the dance demonstrates that revelation, not only transformation, is the goal of liturgical practice. The liturgy does not create an alternate reality but reveals a present truth. This revelation can be seen most clearly in how the church 'frames' its liturgy, the frame inseparable from the performance practice of the liturgy itself. The creation and dancing of the liturgy, the sharing of word and table, is not discovery of a 'higher' truth but the revealing of a present and living truth, leading towards a renewed understanding of the much misconstrued word 'apocalypse'-'revelation'. Revelation is a dimension of performance shared by all ritual, and an understanding of transformation as revelation can enhance our understanding of ritual and perhaps of performance in general.

In recent decades in the United States, the study of performance has turned its critical eye from language to ritual (most notably in the collaborations between performance theoretician Richard Schechner and anthropologist Victor Turner), while simultaneously ignoring the religious implications of many rituals. This raises many issues regarding what it means to perform religion. Is a religious ritual a script or a 'strip of performance' that can be taken out of context and performed for an audience? Can such an action still be called ritual? What makes religious performance religious?

In Nikolas Evreinoff's discussion of theatre as a basic human instinct he writes,

> The art of the theatre is preaesthetic, not aesthetic, for the simple reason that *transformation*, which is after all the essence of all theatrical art, is more primitive and more easily attainable than *formation*, which is the essence of aesthetic arts.
> (Carlson 1996: 35, emphasis in original)

He goes on to say that the ability to imagine realities different from everyday reality and to 'play' with and within these new realities is a condition for religion. 'Man first became an actor, a player; and then came religion' (36). Stepping back from the dead-end of the theatre-religion, chicken-egg conundrum, I offer Evreinoff's definition because I find his characterizations of both theatre and religion to be basically wrong yet widely accepted. First, they portray creativity and imagination in performance as escapist. Religious and performance practitioners alike apparently are not able to deal with reality, so they must create a more livable one. Secondly, privileging formation over transformation obscures the fact that both theatre and ritual do not mask reality in order to overcome it but instead use performance and ritual as tools to unearth, reveal and truly understand the real world. Taking religious ritual out of its context is extremely problematic, which is why understanding the liturgical, theological and sacramental environment of St Gregory's is key to understanding it as performance.

'Liturgy' literally means 'the work of the people.' The meeting, the celebration, the supplication and prayer, the support, the care and the argument that is enacted through this is the work that a congregation does. Liturgy is not a weekly, isolated performance in one space for a set of individuals but a continuous, all-pervading life-orientation that is performed at all times in all places. A church's liturgical celebration is the affirmation that their practice at the altar enacts their engagement with the world and all its people. This is not to suggest that appropriate liturgical practice is easily discovered or that knowledge of liturgical practice is a given. The history of Christian worship is alternately inspiring and embarrassing, as Christians through the ages have struggled to find out who they are through their practice. The real work of liturgy is the discovery of identity through practice, and this identity is continually contested. Writes liturgical theologian Edward Foley,

> Liturgy ... is the bedrock upon which we build our theologies of God, church and salvation. Discovering how these beliefs were embodied and perpetuated by liturgies of the past can help us to look critically at contemporary worship and discover how liturgy is expressing and shaping our faith today. 1991: vii)

In many ways, the fact that this faith is always being shaped and expressed newly and differently is foundational to modern Christians' understanding of themselves as responsive to and responsible for the changing world around them. Their identity is as much in direct response to the world as it is expressed through their liturgy.

Liturgy is by nature theological: it questions humanity's relationship to God and the world. It is quite literally an 'orientation' - a bodily stance taken not only by individuals but the by the body of the church, towards God, the world and each other. Liturgy is a vigorous undertaking of spiritual and moral discovery as well as of decision-making and argumentation. 'Liturgical theology' itself asks how 'the Christian meeting, in all its signs and words, says something authentic and reliable about God, and so says something true about ourselves and about our world as they are understood before God' (Lathrop 1998: 3). Liturgical theology, specifically, is 'the elucidation of the meaning of worship', and as theology, more broadly, it is 'the search for words appropriate to the nature of God' (3). Liturgical theology underscores the necessity of practice in order to undertake the theory; without the practice of liturgy, there would be no theology. For some, this is a beautiful analog to Christian life and belief itself, wherein the following of Christ's teachings is necessarily and forcefully tantamount to understanding them.

Liturgical theology should always be understood in the context of a sacramental theology: the sacraments are the specific instances of liturgical practice where the life of the community, the individual and the divine order fuse in one (albeit multifaceted) event: baptism, marriage, confession, Eucharist, anointing of the sick, taking holy orders, or confirmation. Liturgical life is sacramental life. While liturgy is about the work of the people, the sacraments are about the work that God does through the community's liturgical life.

Sacraments are thus understood as divine action: Christ acts through the action of the believer. The dance of the congregation that is echoed and underscored by the dancing saints and Christ in the mural around them is best understood as sacramental action through liturgy. These sacramental rituals are not merely repeated, the everyday is not simply and temporarily transformed for the benefit of a hopeful congregation, but through the dance the ever-present, active force of God is seen, touched, tasted, heard, smelled. The dance should also be understood as God's sacramental action.

I suggest claiming the term 'rhetoric' for the sacramental work that ritual does, not only because it illuminates the argument-making dimension of ritual, but also because 'rhetoric' is about the interdependence of *how* something is said or done and *what* is said or done. In the same way that rhetoric stresses the interdependence of language and meaning, ritual stresses the interdependence of action and belief, faith and practice. Sacramentality, in the liturgical life at St Gregory's and in the relationship between human action and divine action, is the clearest example of this rhetorical structure. Through it we can link the form of communication and the content of what is communicated. Rhetorics are logics that exist in the performative. Rituals are likewise rhetorics that exist in the intersection between knowledge and belief, between human and divine action.

Christian denominations debate the appropriateness of dance in church - is this action 'too human'? Liturgical dance in the twentieth century United States has only recently enjoyed a revival. The influence of modernity and the mind-body split has taken its toll as liturgical developments and dance followed a similar trajectory by emphasizing the specular and the spoken over the somatic and kinesthetic. For much of the Christian community the *Constitution on the Sacred Liturgy* issued by the Vatican in 1963 was a clarion call, as the Catholic Church sought to reinvigorate its liturgical life by looking to the practices of the people. Although there is no specific reference to 'dance' *per se* in the *Constitution*, Article 30 encourages participation 'by means of acclamations, responses, psalmody, antiphons, songs, as well as by actions, gestures and bodily attitudes,' as well as appropriate times for silence. While some liturgical practitioners focused on 'actions, gestures and bodily attitudes', dance as part of the sacred liturgy only very slowly has started to become accepted within churches of the 'high liturgy' (i.e., Roman Catholic, Episcopalian, some Lutheran). More charismatic churches, which already have a tradition of movement, have been quicker to adopt dance as an expression of faith. Dance as prayer versus dance as distraction and bodily temptation is still a hotly debated subject in Christian churches. However, liturgical dance seems soundly ensconced in some form in Christian worship today, as any brief foray onto the Internet will prove by yielding a plethora of sites offering classes, workshops, costumes and gear, music, books, liturgical dance teams who will perform for or work with your church, and liturgical dance companies who give concerts.

It seems that liturgical dance, whether embraced or frowned upon, is part of the Christian cultural habitus. The dance is a metaphor that has been learned in the body, projected outwards and passed down through generations. Following Pierre Bourdieu, habitus is

> a product of history, [that] produces individual and collective practices - more history - in accordance with the schemes generated by history. It ensures the active presence of past experiences, which, deposited in each organism in the form of schemes of perception, thought and action, tend to guarantee the 'correctness' of practices and their constancy over time, more reliably than all formal rules and explicit norms. (1990: 54)

Even though liturgical dance feels 'new' to many church-goers, the recent reprisal of dance in worship is arguably a 'flowering' of the habitus instituted by tradition within the church. Liturgical dance may 'feel right' to many believers because it is the expression of a deeply ingrained, historically developed attitude towards the world and God. At the same time, discomfort with dance is also a product of another side of that habitus, one that would shun the body and the senses. Even this rejection of dance, as a part of habitus, is the product of a lived, embodied history.

The congregational dance at St Gregory's directly refutes any notion of a disembodied soul or a sinful body, while advocating a holistic spirituality. The dance forces recognition of environment and community in an immediate and necessary way. It creates a living space, what some might call 'flow', the experience that the dance resonates with others, with the earth, with the swing of the galaxy or the tide. This is what American modern dance innovator Ruth St Denis would call 'radiation'. It is an awareness, a presence; it is patience, honesty, openness; an attitude of awaiting more, of seeking transcendence. 'In its deepest sense, all movement in rhythm is religious, revealing in the body the cosmic eternal rhythms of the universe' (St Denis 1997: 73).

The congregational dance of St Gregory's is a lived, embodied metaphor through which the members employ an embodied reasoning to understand and shape their world. Their practice affirms the embodiedness of the mind and the mindness of the body. 'The properties of mind are not purely mental: they are shaped in crucial ways by the body and brain and how the body can function in everyday life. The embodied mind is thus very much of this world' (Johnson and Lakoff 1999: 565). Johnson and Lakoff in *Philosophy in the Flesh* describe how the conscious thinking that we do only takes up about five per cent of all our thought, and that could be a huge over-estimation. In the same way that our bodies function thanks to the unconscious work of the circulatory, digestive and nervous systems, our grasp of complex mental activities such as speech, logic, etc. is made possible by the cognitive unconscious. All that information and perception that makes up the other ninety-five per cent of our thought processes is mediated through the body. When St Gregory's dances, they are informing their lives in ways perhaps not immediately, rationally articulable, but in ways that will be integrated and felt.

As embodied metaphor, liturgical dance can be understood as a cognitive tool by which the congregation conceptualizes. Our physical experiences directly shape how we 'think through' or reason. For example, 'think through' employs a conceptual metaphor of travelling. 'We acquire a large system of primary metaphors automatically and unconsciously simply by functioning in the most ordinary of ways in the everyday world from our earliest years' (47). The metaphors that St Gregory's employ are dance as community, dance as prayer and dance as friendship with God. Not only does the metaphor 'dance as community' become embodied through

© David Sanger / www.davidsanger.com

the practice of the congregation, it serves as an embodied experience of community with the saints, with Christ and with God. The specific actions (circling the altar, singing, making music, touching one another) and the sacred environment (the worship space, the time, the elements such as altar, candles, vestments) are the rites of prayer. The dance is a conversation, wherein the congregants' being-in-the-world' through the dance is a dialogue with God.

In addition, this practice exemplifies a religious life where practice and faith are one. The two intimate partners in this dance are response and responsibility. The dance is a response to God's faithfulness and love that comes through the goodness of the world, the 'many blessings which we bless' when the congregation gives thanks at the Eucharistic table. It is also a responsibility: the congregation holds up their covenant with God by giving thanks through the dance, and the dance is an embodied metaphor for the congregation's covenantal relationship with those outside their immediate community, especially the poor and homeless. Their response is their responsibility in this world-view, which they enact concretely in their ministries that reach into the larger community, such as their food pantry, support of churches in El Salvador, hospital chaplaincy (among several other modes of community service), and in their open posture of 'all are welcome.' Belonging to the community means service, just as faith means practice, and vice versa.

The argument that St Gregory's liturgy of the dance makes is that revelation, not only transformation, is the object of liturgical practice. The congregants are not transformed into better, Godlier people through their practice. Their community is not transformed from a gathering of disparate individuals into a unified body. Rather, the liturgy reveals the truth of the present reality that human community is always and already one with each other, God and the world. Furthermore, it is possible to reconcile these theological assertions with the interpretation and analysis of performance theory. Although the congregation repeats a 'strip of behaviour' in a performance that surely is what Richard Schechner calls 'twice-behaved behaviour', the people are in no way separate and distant from the action of the liturgical ritual. Schechner sees performance as either in the subjunctive mood - looking towards Stanislavski's 'magic if' - or in the indicative, dealing with the actual and historical. In Schechner's model, performance proceeds from the subjunctive, a wish, a myth, legend, into the indicative, reality. Or, performance could proceed from the indicative to the subjunctive, such as in a fictionalized account of a real past event. Other indicative/subjunctive combinations are possible. (Schechner 1987: 38-9) The question liturgical theology posits is, Why should the indicative and the subjunctive be thus held apart, as if they can only work in succession, not simultaneously? There is something lost in looking at ritual merely as a wish. There is something lost in looking at ritual as simply an authentic replication. Rituals can reveal *present* realities, not only ones that must be recaptured and restored, or wished for and created. Together, the subjunctive and the indicative can reveal a present truth that argues for not only a future reality but a radical understanding of the now.

While Schechner's restored behaviour 'offers to both individuals and groups the chance to rebecome what they once were - or even, and most often, to rebecome what they never were but wish to have been or wish to become' (38), the liturgy at St Gregory's is a revelation of the power of the present. This is not to say that rituals do not change over time, or that they are not humanly constructed things, or that the reality revealed is a static, unchanging state. Looking at a Christian liturgy, like any performance or ritual with a past, is gazing at a palimpsest: 'The performance process is a continuous rejecting and replacing. Long-running shows - and certainly rituals are these - are not dead repetitions but continuous erasings and superimposings' (120). Still, the parchment or papyrus on which centuries of scribes impress their marks must offer a kind of consistency, and the ability of the community to understand itself coherently despite this continuous flux is the basis for revelation. Behaviour may be restored, identities may rebecome, but the ability to situate identity sensically within an ever-shifting universe is the radical reality that St Gregory's danced liturgy reveals. All rituals and performances share an element of this revelation, because all rituals and performances are based on an ability to be understood and shared.

Revelation does not necessarily ensure complete understanding. Christian theology's emphasis on the mystery of God affirms that there will always be a place between symbol, metaphor and meaning that may be touched by human understanding but never fully explicated. It is a place of knowing yet not knowing, and this is the space in which liturgy and ritual happen. The liturgy can evolve over time, and therefore its meaning can change, because it is a mode of experience, not a model for experience. This mode is like the space between a metonym and its referent. The sacred dance is a metonym for the life of the church in the world, which is also the 'kingdom of God'. The world will change and the liturgy will evolve, and throughout this process the relationship between human and divine, that radical reality, will reveal itself continually.

Diana Taylor asserts that performances can become so ingrained in a culture that they cease to call attention to themselves as conscious performances. As such, 'scenarios' (such as the 'scenario of discovery' applied by European explorers to the New World) become 'acts of transfer' that transport people, lands or things into the frame of the scenario without regard of right or permission (Taylor 2007: 53-4). The scenario is also a useful tool for critiquing the appropriateness of a ritual or liturgy within the context of the church and world. For however much the liturgy reveals through its frame of practice a world that is the kingdom of God, St Gregory's is also aware that this kingdom is both 'already' and 'not yet', known and yet unknown. That is why the liturgy is the 'work of the people'. They are called to bring that kingdom into being, as well as to participate in it. The gospels can be read as primarily scenarios of liberation that Christ's followers are called to activate in privileging the poor and oppressed. Interpreting the liturgy as a scenario allows participants to critique it as a conscious performance continually in need of revision and work. The liturgy as a framing process reveals that the frame's inclusivity must work for and support those who find themselves outside the secular world's frame of values.

I set out in this paper to understand liturgical congregational dance as theory-making practice danced out in theory-making space, and to place this practice within its historical and theological context in order to comprehend it within the larger context of performance. In my visits to St Gregory's and participation in their dancing, I found that this congregation employs a 'rhetoric of ritual' that reveals their 'theory' of divine community as a living, thriving and present reality. This theory asserts that humanity and divinity share in the creation and care of the world, that the body and the spirit are one just as the practice of this liturgical life is one and the same as responding to Christ's call to community.

What this discovery leads me towards is a conviction that revelation is a quality shared by all ritual and all performance. The word 'transformation' is so easily applied to performance because that is what we immediately see: change. But I challenge the over-use of this word 'transformation' and the assumption that we know what it means. It is so difficult to bring the study of religion into the study of performance because in order to truly understand religion one must step into the theological assumptions the religion makes, and one must make the effort to slip into the skin of 'habitus' that has shaped that religion over time. The possibility of such a thing is highly questionable. It is a question of allowing practice to become faith, something that is viewed skeptically in academic circles. This is why a strictly ethnographic or anthropological approach will always fall short. It focuses on the immediately comprehensible transformations performed in ritual rather than on the revelatory currents coursing deeply beneath the action. On the surface, a visitor to St Gregory's would witness a transformation: the congregation transforms from a collection of individuals into one dancing body, and then after this liminal time of 'communitas', they transform back into individuals and go their separate ways. But underneath and alongside this ritual transformation is an affirmation of what is already embraced, known and understood as a way of life. The congregational liturgical dance at St Gregory's is a transformation, but it is also a revelation, a lifting of the veil to expose an abiding truth. With any kind of performance, whether it be religious or secular, subjunctive or indicative, traditional or avant-garde, transformation as revelation calls me as the performance theorist to invest myself in the world-view and lived reality of my subject in a responsive and responsible way, perhaps in this transformation as revelation uncovering the undercurrents of my own deeply held beliefs while dancing upon the common ground of this beautiful yet needy, nourishing yet hungry - this inexplicable yet knowable world.

ACKNOWLEDGEMENT

To learn more about St Gregory's, please visit www.saintgregorys.org. Great thanks to Dr. Barbara Sellers-Young, whose critical insight helped guide the writing of this paper during the winter of 2007 at the University of California, Davis.

REFERENCES

Bourdieu, Pierre (1990) *The Logic of Practice*, Stanford: Stanford University Press.

Carlson, Marvin (1996) *Performance: A Critical Introduction*, New York and London: Routledge.

Catholic Church. 25 February 2008 <www.vatican.va/archive/catechism/p2s1c1a1.htm#II>

Foley, Edward (1991) *From Age to Age: How Christians Have Celebrated the Eucharist*, Chicago: Liturgy Training Publications.

Johnson, Mark and Lakoff, George (1999) *Philosophy in the Flesh: The Embodied Mind and Its Challenge to Western Thought*, New York: Basic Books.

Lathrop, Gordon (1998) *Holy Things: A Liturgical Theology*, Minneapolis: Fortress Press.

Schechner, Richard (1987) *Between Theater and Anthropology*, Philadelphia: University of Pennsylvania Press.

St Denis, Ruth (1997) *Wisdom Comes Dancing: Selected Writings on Dance, Spirituality and the Body*, Seattle: PeaceWorks.

Taylor, Diana (2007) *The Archive and the Repertoire: Performing Cultural Memory in the Americas*, Durham and London: Duke University Press.

Avant-garde Poetry's 'New Spirit' between Text and Performance, or:
Congregating to disperse and re-member

ROMANA HUK

The artificial border we crossed into a new century marked, in 'post-postmodernist' poetries, a new willingness to reassess the old line drawn between secular and - for lack of a better word, given our underdeveloped language to address such things - 'spiritually'-investigative avant-garde poetics. A helpful prompt to thought, that border nonetheless is dissolving as it becomes everywhere apparent that, as John D. Caputo and Michael J. Scanlon put it in their groundbreaking book *God, The Gift, and Postmodernism*, the last twenty years or more have constituted 'a moment in which the overcoming of metaphysics characteristic of continental philosophy since Heidegger and questions of a profoundly religious character have become increasingly and surprisingly convergent' (1999: 4). Why the convergence? Though I haven't time to do more than gesture, below, at my own narrative of philosophy's return to metaphysics - its understanding, stoked by arguments with postmodern theologians, that its primary heuristic mode, deconstruction, was itself 'structured like a religion', a negative theology of the unattainable 'Other' - it suffices to note that both lost faith in their absolutist practices. And therefore both, through doubt, found themselves opened out to what Jacques Derrida in the 1990s would call 'the possible impossible' just beyond, perhaps, their own discursive certainties.[1] How can doubt, rather than faith, effect 'congregation' around questions decidedly metaphysical in nature? Can there be response to a call to gather that doesn't unify persons around some *pre*text, as of old, but rather an aftertext that they might 'believe' in? Perhaps even one yet unfinished, dependent upon their movements away from it, their dispersal? Why do concepts of performance writing tend to be invoked by poets investigating such possibilities, or: How do lines drawn in the sand - in (millennial) time or on the page - 'get' us anywhere?

For those of us researching the overtly spiritual projects suddenly opening up in poetic practice, the very fact that our queries tend to take the form of Zen koan-like questions, or destabilizing paradox, is itself a provocative lead that simultaneously connects new writing back to earlier avant-gardes, illuminating a 'tradition' for newly emergent work. Many reasons have been given for the hospitality shown, particularly among US artists and poets, to Buddhism and other forms of far eastern spirituality after the door had been pretty firmly closed to most others; political ones are most often cited (and these are certainly relevant to any current conversation about changing forms of sociality and 'religion').[2] But hindsight suggests that such choices also signalled (as indeed koans, too, always do[3]) not impasse for western religious traditions but their enabling breakdown - their *décréation*, as Simone Weil called it, or their deconstruction, as Derrida called it - which was required before another kind of convergence might begin: this time between faith traditions as they would, by the 1990s, become active again in experimental work. Buddhist modes of interruptive thought remain generative today for

[1] These phrases, swiftly becoming commonplace, are again found passim in Caputo's and Scanlon's book, which includes dialogues and round-table discussions between Derrida and other thinkers like Jean-Luc Marion, the well-known Catholic postmodern theologian.

[2] Jennie Klein, contributing several years ago to PAJ's special series of essays and dialogues about 'Art as Spiritual Practice', discusses this phenomenon while reviewing *Awake: Art, Buddhism and the Dimensions of Consciousness*, a multi-venue project devoted to exploring the intersection of Buddhism and contemporary art on the US west coast. Her assertion that '[a]s in the sixties, Buddhism, along with other forms of Eastern spirituality, has become increasingly relevant to many people who see how both Islam and Christianity are used to justify the invasion of other countries' (85) may be somewhat oversimplifying, but it does suggest the connection many made, and continue to make, between absolutist thinking and various forms of aggression.

Performance Research 13(3), pp.109-117 © Taylor & Francis Ltd 2008
DOI: 10.1080/13528160902380

spiritually-inclined writers of all descriptions, because faith traditions too have, like philosophy, come to value *continuing* confrontation with their historical opposites in religious terms. And in the most radical new work, instead of reactionary pursuits of 'visionary' reconciliation or romantic identification with or between them, what continues to be valued are their *differences*, the ways in which encountering nonidentity can force discomfort, self-deconstruction and construction of the new rather than reproduction of the same. As Weil, an important muse of these new poetries, put it in a jaggedly-moving essay written near the end of her short life (which was met in the middle of the Holocaust), after years of thinking between the discomfiting lines of her Jewish-cum-non-affiliating-Christian spiritual leanings, her philosophical/political/mystical writings and exhausting political activism: 'contradiction' is what we have now, instead of 'signs', and therefore seeing it without turning away, with 'detachment', is the only remaining road to enlightenment. She employs other forms of Buddhist hermeneutics in the piece, among others, and manages, by avoiding traditional, too-easy modes of syncretism, to be in sync with both the Christian existentialism of Heidegger, writing alongside her in the 1930s, *and* with the Jewish messianism of Derrida's late thoughts on divinity as the as-yet unthinkable, the coming impossible:

> The contradictions the mind comes up against - these are the only realities; they are the criterion of the real. There is no contradiction in what is imaginary.
>
> ... ...
>
> When the attention has revealed the contradiction in something on which it is fixed, a kind of loosening takes place. By persevering in this course we attain detachment.
>
> The demonstrable correlation of opposites is an image of the transcendental correlation of contradictories.
>
> ... ...
>
> All true good carries with it conditions which are contradictory and as a consequence is impossible. He who keeps his attention really fixed on this impossibility and acts will do what is good.
>
> (Weil 2005: 259)

Like Heidegger's 'earth' - the material 'real' - that juts up into the readable 'world' we make with language, contradictions mark the limits of what Weil calls 'the imaginary'. Meditative attention to the actual multivalence of particulars, and the willingness to be detached from the familiar ways we contain them (and our selves) with words, are the Buddhist tools she uses against her own deepening involvement with Christian mythology. Her words suggest the disorienting physical effects of such encounters with what can't be marshalled into coherence through the usual means: we 'come up against' it; there is a 'kind of loosening'; 'true good carries with it' a burden whose contradictory composition is beyond understanding. And, intriguingly, the only thing for it is to *act* - act, practice, in order to return good to good. Fanny Howe, perhaps the most important Catholic experimentalist currently writing, takes up Weil's ideas at the heart of one of her most recent books of poems, in a prose poetic section entitled 'Doubt':

> When all the structures granted by common agreement fall away and that 'reliable chain of cause and effect' that Hannah Arendt talks about - breaks - then a person's inner logic also collapses. She moves and sees at the same time, which is terrifying.
>
> Yet strangely it is in this moment that doubt shows itself to be the physical double to belief; it is the quality that nourishes willpower, and the one that is the invisible engine behind every step taken.
>
> Doubt is what allows a single gesture to have a heart. (Howe 2003: 25)

In a very real sense, doubt becomes for Howe, as for Weil, not only the physical descendant of belief but also the pivot of her ethics, her politics. 'For only thinking which has the ability to tolerate uncertainty,' writes Gillian Rose, 'is powerful, that is, non-violent'.[4]

3 It might be useful to hear Trappist monk (and friend of modern avant-garde artists like Ad Reinhardt, Robert Lax and Ian Hamilton Finlay) Thomas Merton explain the koan's radical reversal of philosophical logic in his famous encounter, 'A Christian Looks at Zen': 'Zen uses language against itself to blast out ... preconceptions and to destroy the specious "reality" in our minds so that we can *see directly*. Zen is saying, as Wittgenstein said, "Don't think: Look!" (1992: 412). Important to my purposes below, he also quotes Zen scholar D. T. Suzuki to emphasize that '"Zen teaches nothing; it merely enables us to wake up and become aware. It does not teach, it points"' (413).

4 Michael Heller quotes movingly (if without page numbers) from Rose's *Mourning Becomes the Law* in his own very relevant book about lyric, Jewish poetics and American poetry (2005: 230).

I want to pursue belief's 'physical double' elsewhere, however, through the work of another American poet, Hank Lazer, which is informed more centrally and strategically (rather than thematically) by movement, by performance. A long-respected poet in the US avant-garde community, Lazer has taken to describing himself as a 'Jewish Buddhist agnostic', though up until a dozen years ago the secular climate in the arts dissuaded him from fully 'coming out' as a spiritually-investigative writer. More recently, he has become the most articulate thinker about the relations between Judaism, Buddhism and poststructuralism, three of the most powerful forces behind the 'spirit', as he understands it, of avant-garde American poetry. Locating in Derrida's late writing 'a faith defended against faithfulness, or a spiritual experience that exists apart from religious determinisms' (2008: 335), Lazer argues that deconstruction's 'axiomatic' resistance to totalization in all its forms 'may apply equally well to Buddhism, to mystical Judaism and to innovative poetry' (335-6).

As a Catholic by birth, I'm engaged by his work's differing starting point in Jewish philosophy and mysticism, particularly as so much recent Christian thought has depended on its one-time 'other' for revitalization. What Christian theorists call up most often from Jewish thought are two seemingly contradictory principles. On the one hand, and most familiarly, philosophers like Heidegger's student Emmanuel Levinas have, post-Holocaust, stressed the corrective that their tradition presents, given its focus on ethics, on responsibility to the other and on what is immediate, which houses within it the 'concealed' and unknowable in time. Opposing the headier, more hubristic pursuits of 'the ideal' and 'the transcendent' as they have long powered western philosophy and theology, Levinas has become famous for arguing that in Judaism - which lacks even a name for the divine - 'infinity' is to be found first and foremost in the given, the immediate, the face of the 'neighbour', to which and whom one owes one's infinite care. The other's very presence breaks up self-focused priorities and what he refers to as their 'imperialistic' categories of knowing, or controlling; shattering the ego's defenses opens it out to an unfathomably infinite command to care, our only sign of divinity. Yet on the other hand, and despite everything I've said above about the enabling postwar self-deconstruction of textually based belief, Jewish practice also returns us there, if with a stunning difference. Often called the 'people of the book', modern day Jews continue to return us to texts, writing and interpretation as opposed to belief in final revelation and inalterable truth, as Susan Handelman argues in a study of great importance to Lazer that connects ancient Rabbinic methods of thinking to the work of many influential twentieth-century Jewish theorists, from Freud to Derrida:

> The history of interpretation ... has been equally determined by the schism between Jews and Christians precisely over the issue of proper interpretation of the text. Christianity claimed that it had the final and validating interpretation of the now 'Old' Testament text. The word literally became incarnate. The Rabbinic tradition, by contrast, based itself on the principles of multiple meaning and endless interpretability, maintaining that interpretation and text were not only inseparable, but that interpretation - as opposed to incarnation - was the central divine act. (Handelman 1982: xiv)

Derrida's 'deconstructive religion' might itself be said to have sought to 'undo completely the Greco-Christian tradition of thought and replace it with his notion of *Écriture-Writing*', she continues. That Hellenistic tradition's tendency has been 'to *gather* various meanings *into a one* ... [to move] toward the universal, the general, the univocal' (1982: 33). Judaism promotes a very different kind of 'gathering', or congregation: a paradoxical one of calling constant attention to the text through departure from the text, through midrashic interpretation, a *process* by which the text changes with each reader, grows or proliferates, as pages of the Torah demonstrate in visual terms with their gathering marginal commentary. Mystical thought suggests, in versions earlier even than the Kabbalists', that the Torah's 'every word, indeed, every letter, has seventy aspects, or literally "faces"' - seventy being the traditional number of nations inhabiting the earth (Scholem 1996: 62). Rather than corralling meaning within the univocal, as Christian traditions have come to do, such symbolism advocates finding the singular, the 'face' - whether human or linguistic - to be infinitely productive of new meaning. It might even be said, as Handelman does, and as Lazer does too by quoting her appreciatively, that the inexhaustibility of interpretation approximates the divine itself, for in the Rabbinical tradition 'true being was a God who *speaks* and creates *texts*, and *imitatio deus* was not silent suffering but speaking and interpreting' (Handelman 1982: 4; cf. Lazer 2007).

What fascinates me is that such deeply 'traditional' ideas force Lazer to abandon none of the conceptions of the insistently material text upon which most versions of postmodernist poetics were founded; indeed, Judaism seems to have proleptically suggested that 'we are what we write' (a thought I'll return to in a moment). But his old/new Jewish orientation towards language as infinitely productive (rather than continually suspicious for deferring meaning, for not telling us the univocal truth) subtly and positively revises the avant-garde's at-times absolutist ways of regarding the word. American 'Language' poetry's other major tenet, the one that assumes the death of the subject, that forbids assuming the unified 'presence' or unique self-identity of the writing subject, is also honoured in Lazer's practice, though it too undergoes reinterpretation through his adherence to the concept of being called to offer unique *interpretation*, of being part of a community of interpretation - indeed, of 'being in interpretation' (his midrash on Heidegger). In his most recent prose, he quotes (performing what he calls his own 'Talmudic chain of modulated thinking') Levinas's reading of Martin Buber's reading of Hasidic ideas concerning the duty one has of appreciating one's own difference from others: '"This suggests that the totality of truth is made up of the contributions of multiple persons: the uniqueness of each reaction bearing the secret of the text"' (Lazer 2007; quoted in Ouaknin 1998: 60).Each differing speech act, in other words, bears - or, as in Weil's phrase, 'carries' - the secret or unrepeatable truth its *activity* testifies to, which underwrites its offered 'meaning'. In the language of ancient mysticism, each 'soul' must contribute 'its own particular way of understanding the Torah', whose oral component - 'the sum total of ... all others who have interpreted the text' (Scholem 1996: 47) - completes the written Torah. Ironically, paradoxically, then, Jewish thought's corrective to the dominant western onto-theological tradition of valourizing 'presence' and artistic product retrieves them in another form, a *performative* one: one that denies closure or borders around selfhood or the 'work' but preserves uniqueness in the textual *act*, maintaining, just as in the most recent performance theory, that textuality and being are inextricable from one another.

All of which of course impacts upon the way Lazer constructs his activity as a poet. His recent work brings together ideas of the multivalent letter, unique interpretation and speech act/performance by emphasizing the sounding of words - or as he puts it at the end of his book-length, Pulitzer Prize-nominated, 2005 poem, *The New Spirit*, 'sounding it out as you go' (70), recalling the tactile, 'unique' way in which we all first came to making sense of words seen on the page but not yet able to be picked up, used by us. Indeed, one of his several, evolving ways of delimiting the word 'spirit' for provisional use is a midrash on its ancient meanings - in Hebrew, *ruah*, or breath; later meaning the same in Latin, *spiritus* - which locates it as the invisible effusion of the body that animates words, the text; it is, as he calls it, 'the (empty) hub for the wheel of sound ... an inspiration and an expiration' (2008: 194). He thereby works to renew the viability of a 'lyric', or musical, singular practice, one that pursues a 'phenomenology of spiritual experience' (2006: 2) as it happens between text and performance. But his revisionary, provisional definitions of these loaded phrases do *not*, as both 'lyric poetry' and 'spiritual experience' have long been expected to do, summon up a journey of inwardness, of age-old retreat from the world into ecstasy, 'private' revelation and connection to the divine. Lazer's poetic practice has instead become, as he calls it, 'a poetics of *outwardness*': a page-launched practice involving 'an avowedly Talmudic process ... of commentary and adaptation' (Lazer 2007), moving out from texts into other texts, gathering them to depart from them, contributing *himself* as text so that another may ultimately gather *him* in. '[E]verything in the world exists in order to end up as a book', as he puts it, appreciatively quoting Mallarmé (2008: 245),[5] yet extending his meaning through his own tradition to suggest the ultimate Book, the still-unfolding oral Torah, whose unrepeatable 'letters' we both contribute and are.

Bringing all these textual prompts - religious traditional, philosophical, art theoretical and, as we'll see in a moment, social/ethical - together in a new book of prose, *Lyric and Spirit: Selected Essays 1996-2008*, Lazer revises 'thinking as poetry' in the manner Heidegger advocated by making that thinking more communal in project as well as the opposite, paradoxically: more individualized, by opting to embody it, to move towards writing for performance - or documentation, perhaps, of performative being.[6] 'It's not that a poet has something in mind, and tries to express it', he explains, but rather that 'the poem *is* thinking, is an embodiment, a highly specific incarnation and manifestation of an interval of thought' in language (2008: 188). 'The poetic character of thinking is / still veiled over', wrote Heidegger, suggesting that its non-absolutist methods might disclose to greater, as-yet unfully understood, aggregate 'Being' 'the whereabouts of its actual presence'.[7] As if offering an update on both Gerard Manley Hopkins and Heidegger, Lazer gives us a model of the human itself as that which not only seeks to capture the 'inscape' of things, which 'lets things be' to reveal Being, but which *is* precisely because it reanimates or revises its textual world. '*[L]eeshma*   the lyric' (2005: 14), is his suggestive identification in *The New Spirit*; meaning 'for its own sake' in Hebrew, *leeshma*, through such collocation, suggests that lyric-making, sounding it out as we go, is what we do for its own sake and even despite ourselves as poetic thinkers, constantly making new temporary borders to thought, new *art*ificial 'lines' in time. Writing/being involves viewing the spaces of the page as provisional architectures, changing places to move and evolve through. Yet the fact that Lazer's writing has become increasingly attuned, too, to the physical whereabouts, the actual or even regional sites for writing-as-being has a few further implications for his conception of 'congregation' - its politics, its sociality - that I want to end with, through looking at the work itself.

II.

The New Spirit is Lazer's most extended experiment not only with contributing a 'phenomenology of spiritual experience', particularly as pursued through 'the possibilities of musicality and thinking / singing' (2008: 193), but also with situating his work - by attending to, as that other tenet of Judaism demands it, 'the neighbour', the other. This involves hearing, first, as Lazer suggests in a recent essay about 'the metaphysics of sound' that dwells for a moment on the beginning and impetus for the poem:

> *The New Spirit* provides the room for writing that is both a thinking about musicality and a thinking within musicality (or an embodiment of it). From the very beginning of the book ('Prayer'):
>
> any one   could be the one   the sudden
> stun   you'd waited for
>   arrest again
> a rest   against the elements
>
> The book-length poem focuses on how we hear and considers spiritual experience as something that

[5] The quote is from Mallarmé's 'Le Livre, Instrument Spirituel' (1996: 14-20).

[6] See Auslander for his recent thoughts on 'the performativity of documentation itself' (2006: 5). Lazer's work even more radically aligns writing and being. As I use the phrase 'performance writing', I assume that readers of *Performance Research* will understand I use definitions developed most clearly by Caroline Bergvall and the programme by that name that she directed in the 1990s at Dartington College of Arts. Bergvall has had a profound effect upon the American poetics scene, which until very recently - and texts like Charles Bernstein's edited volume *Close Listening: Poetry and the Performed Word* (1998) - had comparatively little interest in performance beyond the 'poetry reading', which is a very different thing. The idea, as Bergvall put it at the programme's first symposium, that 'everything about a [written] work is active and carries meaning', rhymes very powerfully with Lazer's Jewish interpretation of texts and his own writing practice. His growing interest in writing and spatial architectures is also deeply in sync with Bergvall's work.

[7] See Heidegger's opening, poem-like statements entitled 'The Thinker as Poet' in his later work, *Poetry, Language, Thought.*

occurs in large part through a summoning to hear (as in the central Hebrew prayer, the Shema - 'Hear O Israel ... '). (2008: 193, Lazer's final ellipsis)

Like the text, to which one responds in order to be, to which one has the responsibility to respond, 'any one' might deliver the 'stun' of otherness that 'you'd waited for' - that delivers you up, paradoxically, by also demanding response that momentarily halts the flow of singular being. The (several times) daily prayer Lazer refers to after the quote continues: 'The Lord is our God, the Lord is one'; in Lazer's interpretation, any 'one' might be that deliverer. The openness to otherness that initiates this work might be said to be like the response that, Levinas argued, reduces the ego to the 'here I am', Abraham's '*hineni*' in response to the unfathomable Other who calls him to do something - something he doesn't fully understand. The Hebrew word translates as neither 'behold' nor 'here' but rather something in between, a gesture, 'the speech act of making myself available to another' (Putnam 2002: 38). Revising conventional notions of 'Prayer', this first section of the poem begins not with address to a deity but a turning to the immediate, finding its start in a Levinasian moment of enabling 'stun', of encounter with one's 'neighbour'. Such an 'arrest' (scored throughout the poem by spaces between phrases) calls for response, an alteration of direction - even a kind of enabling death to self, as happens in Abraham's story, and as happens next on this page as 'persephone personifies the dying into / dynasty' (Lazer 2005: 9). This textual fragment of Greek creation story (which ironically links the beginning of time to death, to Persephone's renewing date to be kept with Pluto in Hades) will come to locate the poet's whereabouts on a number of levels, as he reflects on his father's recent death and his own turn to mid-life, and as he joins time's unfinished text by writing the mortal body through 'sacred expiration' (12). But here, the focus is on the unmistakable pleasure, too, of a hearing that encounters the 'stun' it's waited for, like a would-be lover, that opens it up, allows in proximate difference - all of which happens throughout the poem in sound: 'arrest again' beside 'a rest against', and 'dying' hearing its kinship and differences with long-lived 'dynasty'. Writing lyrically is an act of hearing - especially hearing that which contradicts one's own 'line' and prompts new movement that speaks to difference.

It could be 'any one' 'stun' that gets the poet's writing moving - so does it matter which? In the years just preceding the writing of this work, Lazer had begun theorizing the importance of regional location in experimental poetics, despite our 'experience [of] the contradictory impulses of a digitally accelerated movement towards globalism' (2003: xv). Such issues become significant to the investigation of any phenomenology of spirituality as well, '[b]ecause of course "soul" too has a history; it is not a transcendental category apart from time, the contingent, the material, and the social ... [I]t too is a socially and culturally mediated act of imagination' (2008: 67) - as writers in a diasporic tradition know far better than most. As Handelman puts it, connecting Jewish textual practice, its midrashic unfolding, to its historical circumstances: 'Displacement is a necessary re-vision and re-creation of a text which is the only anchor of a people displaced in space. Displacement, in other words, is both the *condition* of and the *answer* to exile' (1982: 223). Lazer's movement years ago from his birthplace on the US west coast - where his grandparents had migrated, fleeing the Russian pogroms - to Alabama, deep in the American South, can hardly be seen in terms of 'exile'. But it does seem to have produced a response in this work, one that brings his own diasporic tradition into differential relation with that other once-enslaved diaspora whose historical experience in the region remains so palpable there - and perhaps most poignantly in Alabama, the site of the start of the Civil Rights Movement. The very second page of his long poem picks up strains of this 'neighbouring' tradition (perhaps as emanating from that major port for importation of slaves, New Orleans, where French, Spanish and young American rule succeeded one another, and where those great African American contributions to the arts, the blues and jazz, would flourish). What he hears is a 'singing *in* the telling', both like and not like that of his own lyric mode, perhaps because its highly 'unique' embodiment of historical experience in articulation had no choice but to sound out, reinterpret, available (and allowable) words and modes of speaking, singing, chanting, praying:[8]

> *tristesse tristesse*
> my brothers bless sold away sold away
> sweet soul toil by day sold away sold away
>
> *'so to tell and not to sing it'* the singing *in* the telling
> syllable by syllable held the echoing way toil by day
> sweet soul doing as you are told sold away
>
> *tristeza* bless my brother's head
> blues implode
> hold him holy bless my brother let him live
>
> we think in fact we think we hear so long so long
> the silence couldn't last long lost lyric his
> body torn apart *tristeza* bless my brother's head
> (2005: 10)

This most sonically balanced (and perhaps traditionally 'beautiful') page in the book gathers into its inaugural section called 'Prayer' a rhythm of language use that will inform the poet's own and the whole composition, returning to be 'held [in] the echoing way' of the writing's hearing/thinking/making. Though on the next page he hears potential critique, too, of his possibly too-reconstructive, anachronistic retrieval - 'no more than a postmodern did you say / post mortem period piece in the authentic rhetoric' (11) - his insistence will be that there is 'no end to what begins' (12), what is made/heard, and that all hearing is a 'beckoned listening ... / summoned from the criss-cross of your historical circumstances' (16). '[W]hat does change' is only 'which combinations of sounds what music one / in his or her time is inclined to listen / to' (16). Listening takes on an inescapably political as well as ethical dimension, as does the form of one's response. For a poet whose post-Cage, newly constructivist poetics 'call[s] the language place / the house of being a dwelling place for time & the life / of consciousness in it / tenant in it / subject / to eviction' (60), what is heard/written is capable either of changing the direction of one's (and thus everyone's) unfolding text towards a fuller congregation, a fuller playing out of difference, or of 'evicting' from it those most subject to its willful closure. Though no one can, obviously, write all, not 'Leaning Toward' and hearing what is most immediate, as the poem's last section title has it, is all that might threaten exile for others.

And it is through such leanings and turnings, 'stuns' that take one off course, that *The New Spirit* moves and suggests what its own title might be moving towards meaning. The poem's third of four sections is named '*Teshuvah*: Heading South', reflecting (in part) the poet's physical, literal directions in time, and his move into new geographical 'circumstances'. The word in Hebrew is a multi-layered one that signals a journey of the spirit, which is what, of course, this poem also is, as a whole. But more specifically, the word suggests 'reply', as well as 'turn' or 'return', and even 'repentance' - becoming the word for the period of atonement that traditionally culminates in Yom Kippur. Lazer of course reinterprets of all of this to offer its meaning for him: '*teshuvah* turn toward' (46). For him, the period of self-reflection that is Teshuvah leads not to reconciliation with traditional belief but an ability to turn once more to otherness - through a self-deconstructive but profoundly *communally* re-constructive, performative/physical 'grasping of self as grasping of non-self' - which is Derrida's definition of 'spirit' (Lazer 2008: 226; Derrida 1989: 26).This may seem to contradict the idea of contributing the 'unique' writing - but does it? Lazer's conception of the spirit or 'soul' would suggest, as he does by again quoting Derrida,

[8] As Norman Fischer - American avant-garde poet and Zen Buddhist priest - explains, 'Since African captives were not allowed to use their native music in America, they created, out of American folk and church singing, and the conditions of their own lives, a new art, their own saving way of expression (1993: 44).

that it is 'that which gathers or in which what *gathers* is gathered' (Lazer 2008: 238; Derrida 1987: 76) - i.e., 'spirit seems to designate, beyond a deconstruction, the very resource for any deconstruction' (Lazer 2008: 226; Derrida 1989: 15). It does so by being, paradoxically, as I suggested at the start of this essay, the very energy, inspiration, that welcomes otherness in, welcomes self-breakdown, feels the drive to put oneself in service of the other:

traction   your face
                    immediately places me
            in time   see   hear
                        location circulates

soul too has   its own autonomy   & can be felt
taking its own way   a surprise to my thinking
    teshuvah   soul
turns quietly toward   (2005: 61)

Creating a whole new meaning for the old line 'the spirit moved me', Lazer makes uncertain, 'surprising', movement itself the 'heart' (as Howe put it) and soul of his 'practice' - a word I want to use here, in conclusion, for all its many meanings.

The last two sections of *The New Spirit*, 'Teshuvah: Heading South' and 'Leaning Toward', offer lyrical turns to, interpretations of and extensions of texts/sounds like the age-old 'When the Saints Go Marching In' - that gospel hymn, originally a 'spiritual', that began its life as a funeral march and has, over time, been recorded again and again, by dozens of jazz musicians and pop artists, until it seems there is no single possible 'official' rendition of the piece. Remembering it as what it was at the start - in New Orleans, known as a 'jazz funeral' - Lazer brings it in as accompaniment here to his work's own 'sacred expiration', its giving over to otherness, its dying into living text. Re-membering it here, alongside many others, many new renditions of it (like the poet John Taggart's) takes his thought somewhere by improvisation, listening, to a place not yet known: '[t]he musical space entered into in the composition of *The New Spirit* is "not a known condition", just as "toward / the middle   [Col]trane played ahead of any sense he understood"' 2008: 203; 2005: 35). But just as dirge may become, through some interpretations, wildly celebratory music, such radical openness of making prophesies the bright potential for continual, congregative construction built, paradoxically, on hearing '*difference   difference   difference* / the common / denominator'(2005: 23).

> For the world we live in and our particular singular instance of being, we have the opportunity to take a sounding, to learn to listen to the sounds of our time, and to offer up our thinking/singing, a particular and idiosyncratic sounding that takes its part in a larger and often unapparent choral offering, a collective that we participate in by virtue of our peculiar human residency in and determination by language. The sounding by means of poetry is perhaps our best and most serious play, our playing with the instrument and exploring the possibilities of and in language. (2008: 204)

'And thus perhaps collectively, and unknowingly', as he put it in his introduction to a collection of works from his own particular place in time, 'we - the poets of *Another South* - are together assembling a vast (verbal) Paradise Garden' (2008: xx).

REFERENCES

Auslander, Philip (2006) 'The Performativity of Performance Documentation', *Performing Arts Journal* 84: 1-10.

Bergvall, Caroline (1996) 'What do we mean by performance writing?', keynote address at the first Performance Writing Symposium, Dartington College of Arts (currently unpublished manuscript).

Bernstein, Charles, ed. (1998) *Close Listening: Poetry and the Performed Word*, New York and Oxford: Oxford University Press.

Caputo, John D. and Scanlon, Michael J., eds (1999) *God, The Gift, and Postmodernism*, Bloomington and Indianapolis: Indiana University Press.

Derrida, Jacques (1989) *Of Spirit: Heidegger and the Question*, trans. Geoffrey Bennington and Rachel Bowlby, Chicago: University of Chicago Press.

Fischer, Norman (1993) 'Buddhism, Racism and Jazz', *Tricycle: The Buddhist Review* (Summer): 41-5.

Handelman, Susan A. (1982) *The Slayers of Moses: The Emergence of Rabbinic Interpretation in Modern Literary Theory*, Albany: State University of New York Press.

Heidegger, Martin (1971) *Poetry, Language, Thought*, trans. Albert Hofstadter, New York: Harper and Row.

Heller, Michael (2005) *Uncertain Poetries: Selected Essays on Poets, Poetry and Poetics*, Cambridge: Salt Publishing.

Howe, Fanny (2003) *Gone*, Berkeley and London: University of California Press.

Klein, Jennie (2005) 'Being Mindful: West Coast Reflections on Buddhism and Art', *Performing Arts Journal* 79: 82-90.

Hank Lazer (2003) 'Introduction', in Bill Lavender (ed.) *Another South: Experimental Writing in the South*, Tuscaloosa and London: The University of Alabama Press.

Lazer, Hank (2005) *The New Spirit*, San Diego: Singing Horse Press.

Lazer, Hank (2006) 'Hank Lazer Interview' by Jeffrey Side, *The Argotist Online* (posted 2 May) <http://www.argotistonline.co.uk/Lazer%20interview.htm>.

Lazer, Hank (2007) 'Is There a Distinctive Jewish Poetics? Several? Many? Is There Any Question?', Annual Duffy Lecture delivered at the University of Notre Dame, 14 November. Scheduled to appear in the winter 2009 issue of *Shofar*.

Lazer, Hank (2008) *Lyric and Spirit: Selected Essays 1996-2008*, Richmond, California: Omnidawn Publishing.

Mallarmé, Stéphane (1996) 'Le Livre, Instrument Spirituel', in Rothenberg and Guss (eds) *The Book, Spiritual Instrument*, trans. Michael Gibbs, New York: Granary Books.

Marranca, Bonnie (2002) 'Art as Spiritual Practice: Panel Discussion with Alison Knowles, Eleanor Heartney, Meredith Monk, Linda Montano and Erik Ehn; moderated by Bonnie Marranca', *Performing Arts Journal* 72: 18-34.

Merton, Thomas (1992) 'A Christian Looks at Zen', in Lawrence Cunningham (ed.) *Thomas Merton: Spiritual Master*, New York: Paulist Press.

Ouaknin, Marc-Alain (1998) *The Burnt Book: Reading the Talmud*, trans. Llewellyn Brown, Princeton: Princeton University Press.

Putnam, Hilary (2002) 'Levinas and Judaism', in Simon Critchley and Robert Bernasconi (eds) *The Cambridge Companion to Levinas*, Cambridge: Cambridge University Press.

Scholem, Gershom (1996) *On the Kabbalah and Its Symbolism*, New York: Schoken Books.

Sherwood, Kenneth (2006) 'Elaborate Versionings: Characteristics of Emergent Performance in Three Print/Oral/Aural Poets', *Oral Tradition* 21(1): 119-47.

Weil, Simone (2005) 'Contradiction', in Siân Miles (ed.) *Simone Weil: An Anthology*, London: Penguin Classics.

Become
Secret acts in the work of Kate Davis

KIRSTEN NORRIE

'A miniature love story frames the discovery that a certain Boutades, a Sikyonian potter at Corinth, once made: he manufactured the first portrait image on behalf of his daughter, "who was in love with a young man; and she, when he was going abroad, drew in outline on the wall the shadow of his face thrown by a lamp. Her father, having pressed clay onto this, made a relief that he hardened by exposing it to fire along with the rest of his pottery"' (Tarn Steiner 2001: 3)

• **Aphrodite Kallipygos** **2007.** *Pin pricked paper 186 x 82 x 2cm*

PRICKING APHRODITE

In 1946 Marcel Duchamp made a glutinous painting consisting of a daub of offish white on a black background, entitled *Paysage fautif (Harem-Scarum Landscape).* The word play entitling this work became all the more explicit when it later transpired that the substance he used to paint it with was his own semen. Inseminating art work has been a less literal activity for centuries in the West, its metaphoric implication imbuing the activity of applying paint to canvas witnessed exaggeratedly in the work of male artists such as Jackson Pollock; in the vivification of Greek archaic and classical statuary through its mnemonic function to recall those lost at war or at sea; and in the incestuous relationships evolved between artists and their work (Paolo Ucello's wife beleaguering him to quit perspective and join her in bed). Yet what is even more interesting in the case of *Paysage fautif* is not its explicit performativity but the covert privacy in which it was enacted and the inevitable status of documentation it thereafter attained.

Kate Davis's work involves moments of intense private action. As an artist who sculpts, draws, makes video, text and installation, Davis's work has often proved resistant to language, attempts at description frequently caught up in a tautological reworking of events. This is not due to the number of areas she works in, but the deliberately subtle and complex interplay between parts. *come,* shown at Milch Gallery London in 1997, comprised complex works such

Performance Research 13(3), pp.118-122 © Taylor & Francis Ltd 2008
DOI: 10.1080/13528160902819406

as *Negligé (Woman with her Throat Cut)* - thirty-six pin punctured drawings at ten degrees apart transcripting Giacometti's sculpture. Also included in *come* was *Drawing Towers while lying down*, an impressive biro inscribed sheet of glass tilted against the gallery wall. Taking almost three hundred hours to create, the repetitious nature of the work is accentuated knowing Davis's near blindfolded state in drawing through carbon paper. *Little Red* incorporates metal etched glass and the obscured shape of a lemon, abstracted by its lack of colour, cut in half, the symbol of bisexuality due to its hermaphroditic incorporation of both male and female parts. The vaginal appearance of the cross-sectioned fruit is heightened and hidden by Davis' placing it cut surface down; a subtle wound that iterates the tactile quality in much of her work. The nuanced and covert allusion to touch; sometimes a precisely damaged touch in the form of slices, pricks, slight but multiple burns and often a touch that elicits change through repetition has been extended to recent work.

Headhearthole, title of Davis's exhibition out of a 2007 residency at the Wordsworth Trust, Grasmere, included a delicately pricked: *Aphrodite Kallipygos*. The goddess was the same height as the artist. The etched glass of *What?*, a large panel whose spray of metal flecks adhering to the surface corresponded to the shape of an ink blot, evoked comparisons to the lumbar region of the spine or the thorax region of an insect. The piece was infused with the same punitive potential as *Little Red*; Davis held a grinder with her back to the glass and bent over, the steel clasped between her feet, so that sparks shot out between her legs, settling in a scorched spray on the transparent surface.

IN ABSENTIA

'I was brought up to pray as if everything depended on prayer and to work as if everything depended on work. There was no room for anything else . . . I kissed the feet of Christ and sent the statue flying, shattering into a thousand little pieces on the floor. I will never be able to put him back together. I am destined to be on the outside from now on.' *(Some References to red - remembered, conjured or stolen while walking for 28 seconds* Lecture, Kate Davis, Milch 1998.)

Stone statues have long been used in funerary rites as substitutes for the dead. As compensatory doubles in Ancient and Classical Greece, the statues often suffered punishments meted out to them as if to the living embodiment of the person they represented. 'Thaegenes of Thasos, Olympic victor in 480 and 476, was honored with a bronze statue standing in the agora of his native city. Flogged by an enemy after the athlete's death, the statue toppled down on its assailant and, on being convicted of homicide, was thrown into the sea' (Tarn Steiner 2001: 8). The function of wax, lead and clay effigies employed in fifth-century Athenian curse rituals was no less rigorous; figurines mutilated, burnt or bound in an attempt to damage specific and relevant parts of the original person portrayed. The invisibility of the 'original', the animated thinking and feeling flesh of the statue's human counterpart suggests an autonomy and focused visibility all too easily centered on a figurative replica. However, as Socrates suggested, the ideal portrait would consist of an amalgam of human properties, adding that the allusion to the soul should be the central concern in the fashioning of an idealised body. 'The statue maker ought to make the outward form like to the workings (erga) of the soul' (Tarn Steiner 2001: 34).

The metonymic presence of Davis's body is apparent in *Philosophical Object* (1997), for example, which presents the viewer with a cylindrical glass tube, the outside sprayed with a film of vaporized aluminium creating a separation between surfaces. The length is equal to the artist's forearm and the internal diameter is 10 cm - the maximum dilation of the cervix at birth. The sexualised metonymy of this work and *Aphrodite Kallipygos* presents Davis's presence as a subtle mnemonic device, alluded to in the potentially entrenched forearm and a reworking of personified love given the same stature as the artist. These elements are potentially incorporated as contiguity rather than similarity - the artist's body extended through the work in a kind of tangible sympathy creating 'a bond that need not rest on any visible mimetic likeness' (Tarn Steiner 2001: 3), albeit selective likeness. It is this exactitude which can be found in all aspects of Davis's practise, elevating acts performed within her process to defined moments of private performance.

The removal of significant value in order to create something of equal value is a consistent concern within Davis's work, particularly apparent in the drawings made painstakingly with a pin. As an anatomy (anatomising the materiality of the work itself), it rests equivocally between construction and deconstruction enabling a point of equipoise, a resting place, albeit an uncomfortable one. The somatic inference here of mystical or mysterious bodies, latent within *Aphrodite Kallipygos* and other significant works such as *Little Red*, is that the boundaries of knowledge and belief are open to reinvention, particularly when ascertaining anatomical 'truth' in confronting interiority, for example. However, Davis consistently concentrates on the exterior form of her figures, transfixed by Bernini's *Ecstasy of St. Teresa*, Werner Herzog's *Heart of Glass* and Giacometti's *Woman with her Throat Cut* - all of which allow interior states to dictate or shape external appearances. For St. Teresa it is the dichotomy of her climactic rapture, whether sexual or religious; for the cast of *Heart of Glass* it is the hypnosis they undergo to depict the love, obsession and murder surrounding the lost procedure to create the infamous ruby glass; and for the subject of Giacometti's sculpture it is a woman's stance in death. Rather then, the exterior images created by Davis suggest a psycho-sexual quality reverberating with emotional intelligence which is achieved, in part, by a particular handling and positioning of materials. It is this aspect of Davis's practise that encourages a revised reading of the performative act where the object or prop is the culmination of a considered act within process as opposed to the more disposable prop of theatre or the live art object to be utilised as idea/action for the duration of the piece. Davis seems to refute this position by producing permanent evidence of live action purposefully imbued in the materiality of the object itself.

THE CORROBORATION OF TOUCH

Red Shift shown at Milch in 1998 is a video work comprising the artist moving away from the camera, her back towards the viewer in a simultaneous advance and retreat, the scarlet of her coat at first fully saturating the screen then receding as the figure moves away to produce enigmatically sexual shapes against a white background that could be interpreted as snow. Davis is fully aware of the indelible nature of red, an undertone to the seemingly bleached formality of works such as *Alone*, Milch 1998, where despite a tincture in varnishing a lemon with scarlet nail polish, all else remains white - a small table, marble-like folds of large white sheets of paper and a mirror with a white spray painted ellipse. The properties of red apparently bleed through in this delicately psychosomatic piece, but in unexpected places. (It is worth noting that 'alone' in Italian can also mean stain or halo).

Yet the corroboration of touch wilfully denounces accident in the construction and dissemination of such pieces, instead of which the artist has specifically chosen the formal moments and objects to imbue a red-edness. This colouring up of work encourages an emotive position within a minimally formal one, excited by the rigorous possibilities of repositioning Bernini's *Ecstasy of St. Teresa*, for example, and this is true of the succinct acts enmeshing the objects and drawings Davis has recently created with a particular interpretation of the emotive properties of form. It is this knowing detachment that allows the viewer distance from the rawer interface of private, personal and subjective emotional experience. The knowingness of touch, the preconceived relationship to Davis's creation of recent objects and drawings could be said to harbour instinctual intuitive emotion and contemplation; certainly by the conditions of near silence (the pin pricks or flying sparks existing usually as afterword or rumour) and seclusion. But it is the action itself which already bears the implications of feeling; the propensity toward pain, repetition, conscientious action, heightened by a lack of visibility that allows Davis to coolly detach and re-align the intensive history of each work within the pieces she groups together.

OBJECTS OF CONTEMPLATION

'The bishop bends towards me to give me holy communion. I open my mouth to receive the body of Christ and before I shut my eyes I am suddenly faced with an aerial view of an open vagina. The bishops mitre. Red and shaped like the splayed female sex. Blood rushes to my cheeks, glowing with heat I am embarrassed not from what I see but from what I feel' (*Some references to red - remembered, conjured or stolen while walking for 28 seconds* Lecture, Kate Davis, Milch, 1998).

This deference to her materials, to the conscripted act of making, poses a formal question for Davis herself - how can she connect the unmade to the made? In accentuating a vital self-awareness, by subscribing to the process of creation a vigorous and equal element in relation to the finished work, albeit a private one (and this is important), perhaps this inevitable fissure is temporarily overcome. The blindfold of carbon paper in *Drawing Towers while lying down* evokes religious artistic practises where the artist may not look at what they are conveying. Instruction through mirrors, where the object increasingly accrues more awesome/fulness than the individual creating it, is commonplace in the history of religious artefacts. This is deflection rather than peripheral vision, another fact possibly alluded to in Davis's consistent use of mirror, glass, nail polish and hard surfaced metal. The deflection extends to the pin prick drawings with the rhythmic extraction of tiny discs of paper, an elaborate sense of reduction metaphorically and paradoxically analogous to the creation of an icon. The paradoxical tenderness applied in a potentially punitive fashion is reminiscent of the death by a thousand cuts, the gradual blinding of seamstresses forced to work by candlelight in eighteenth- and nineteenth-century Britain and early 'tattowing' with a sharp wooden needle and ink. Reminiscent too of the small penetrations of cupid with his tiny arrow, somehow impotent in the enlarged and life-sized evocation of Davis's Aphrodite. In looking back over her shoulder at the viewer, this figure comprising many holes in the paper evokes the unturning red figure of *Little Red* in an uneasy equivalence mimicking Davis's assertion that 'there are two female fluids. Ancient sources called these the River of Life and the River of Death, meaning the clear or white flow at times when a child is more likely to be conceived and the forbidden red flow of menstruation, when it is most unlikely that a child can be conceived.' Of course Aphrodite was never meant to be fertilised; her calling is for the consummation of eros as a self-contained act. She is punctured into existence through the repeated penetration of a needle which brings her soft flesh into light-suffused being. It is too easy to read red and white as binary opposites; the staunch and the flow are perhaps more symptomatic of the kind of emotional defiance and reverberation suggested in redness as a position. In this light, Aphrodite's punctured form is cloaked in an attitude of red that declares itself through the repeated penetrations and the admonished reduction of the act of love.

Concurrent states of absence, terminally referencing the physical presence of the artist, point mnemonically and dramatically to the death of self, inferred in *Negligé (Woman with Her Throat Cut)*. That the artist must sacrifice a confrontational and holistic portraiture not only of her physicality but of the chosen depictions of emotional, psychological and sexual is only natural as one who traverses 'the outside'. Peculiarly classical in its aims, the rigour and strict adherence to boundary within much of her work does not compromise the bleed of more implicit presences. Tracing the boundaries of conflagrating catholic and personal concerns, the work mimics and shadows the processes, images and practices of the faith; transubstantiation operating as the transformative and shifting meaning within much of her work where the act of making is an act within itself - it becomes.

Integral to this position is deference to substance, for of course her objects remain still, once released from the turbulent and painstaking demands of their creation. And yet it is this stillness that allows us to witness the gamut of attendant feeling associated with the imagery Davis uses. Insofar as all objects have the potential to be used, they are performative.

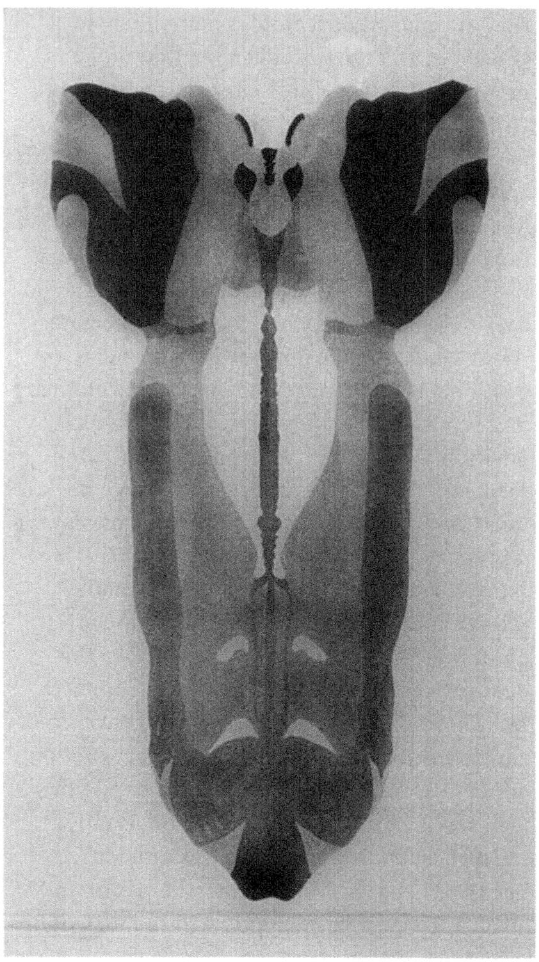

• ***What?* 2007.** *Steel drawing on glass, 196 xx 113 x 25cm*

This is accentuated by the position of interiority, fashioned out of specific intent in shaping and defining her materials to correspond to an interplay of somatic desires and frustrations. The conglomeration of specific parts within Davis's work, some alluding to classicism and perfect form: *Aphrodite Kallipygos*; the polished finger in *Little Red*; the cuts of *Negligé (Woman with her Throat Cut)*, imbued with a consciousness of artistic presence, of authorship sometimes metonymic, other times metaphoric, formally abstract or minimally defined could be read as a congregation of self, the constituent parts meeting in a vivid collusion that references the absence of the artist only to be re-iterated by her suggested presence. The inference in every one of those moments, heightened by the suggestion of private ritual, is of an instinctual system of reduction and construction shrouded in the mystery of oneself in absentia. The ever-departing Davis in *Red Shift* suggests a continuation of red as a condition, a belief through the transformative appearance of the artist at first wearing it and eventually, through distance, becoming it. The distance we are never very sure of, she carries herself with it and as she departs a small evolution occurs where, walking away - aphrodite with her back turned - she becomes a version of red she permits us to see: herself.

REFERENCES

Tarn Steiner, Deborah (2002) *Images in Mind: Statues in Archaic and Classical Greek Literature*, Princeton: Princeton University Press

A Mountain Communion

NATHAN WALKER & ELANOR STANNAGE

ONE OF THE SMALLEST FELLS IN THE LAKE DISTRICT, CASTLE CRAG STANDS IN THE BORROWDALE VALLEY. A LARGE OUTCROP OF STONE PUNCTUATED WITH TREES, THROUGH FORMER INCARNATIONS AS THE LOCATION OF AN ANCIENT FORT, A QUARRY AND BORROWDALE'S MEMORIAL TO THE FALLEN OF THE FIRST WORLD WAR, MORE RECENTLY IT HAS BECOME THE SITE OF AN ANONYMOUS AND ORGANIC CONGREGATION; A COMMUNION BORN OF OUR DISPARATE TIMES. IN THE PLATEAU OF A DISUSED QUARRY, SLATES ARE LIFTED AND PILED. WE VISIT THE SITE SEPARATELY, THREE TIMES. ON OUR FINAL VISIT THE HUNDREDS OF UPTURNED FLAGS OF SLATE ARE FALLEN, THE CHARGE OF MEMORIES REMAINS.

CASTLE CRAG, BORROWDALE, CUMBRIA. ANONYMOUS MEMORIALS APRIL – DEC 2007

ALL IMAGES ARE POLAROID 1200 (EXPIRED) TAKEN ON AN IMAGE SYSTEM CAMERA BY NATHAN WALKER

Performance Research 13(3), p.123 © Taylor & Francis Ltd 2008
DOI: 10.1080/13528160902819414

The Divine Spaces of Metaphysical Spectacle

Ruth St Denis and Denishawn Dance Theatre at Lewisohn Stadium, the esoteric model in American performance

SUSAN TENNERIELLO

> I demand of the dance more than any of the other arts that it reveals the god in man. Not merely the scientific and beautiful forms that his body can be made to assume, but the very divine self. Ruth St Denis ('Credo', 1925)

Over a century has passed since Ruth St Denis (1879-1968) first performed *Radha* in 1906 at the Hudson Theatre in New York. The sensational debut of this modest young woman from rural New Jersey, who aspired to transform stage-dancing into a respected art form, retains the aura of that mythic American dream that promises boundless opportunity and personal freedom. Appropriately, she wanted to be remembered as a pioneer. Although St Denis is best known as enveloping herself in an exotic persona, she exuded the trail-blazing enterprise associated with 'the American spirit'. A self-styled 'interpretative dancer', St Denis cultivated an eclectic, intercultural aesthetic based on gesture and movement exercises developed by the French music teacher François Delsarte, in commercial theatre spectacles, in Christian Science, in theology, Hindu philosophy and an ever-expanding study of Indian, Chinese, Japanese and Southeast Asian theatre and dance forms. First as a soloist and then in partnership with Ted Shawn and their company Denishawn, St Denis's ambition was to ressurrect the sacred function of dance as primary ritual of cultural value.

Denishawn was the first American company to tour Asia (in 1925-6), performing in major cities and provinces throughout Japan, China, Malaysia, Burma (Myanmar), India, Indonesia and the Philippines. On its return from this tour, the company premiered a new programme based on its impressions and study of Burmese *pwe*, Javanese court dance, Peking Opera and Japanese kabuki. After visiting Asia and seeing first-hand the cultures that had inspired her for over twenty years, St Denis often spoke of how the artistic traditions she saw held an integral function in the cultural and religious ideology of each society. She noted, 'as a young nation, we still need to drink at the old fountains. Nature and art, in their highest forms, are the only sustenance fit for the growing child' (1939: 328). St Denis dwelled on the patronage granted other cultural insitutions, such as museums and orchestras, believing that American dance deserved similar recognition. Emboldened by the tour, she claimed that she was finished with 'Oriental' goddesses. However, she was not finished as the *Goddess* of dance. The 'divine dancer' returned to her Christian roots.

Between 1928 and 1931- the year when the Denishawn school and company was disbanded - St Denis began to revise her concept of dance drama with a form she called 'temple ballets'. These dance rituals were inspired by liturgical subjects and metaphysical principles derived from world religions, mystic and estoteric strains of transcendentalism, Christian Science and hyperspace theory. I highlight St Denis's staging of Denishawn dance spectacles *The Lamp* (1928), *Angkor Vat* (1930) and *The Prophetess* (1931), created specifically for appearances seen by

Performance Research 13(3), pp.124-138 © Taylor & Francis Ltd 2008
DOI: 10.1080/13528160902819422

thousands at the Lewisohn Summer Concert Series in New York. Beginning with *The Lamp* and pulsating throughout *Angkor Vat* and *The Prophetess*, St Denis declared: I am the light, I am the guide. In these 'prophetic' dance dramas, St Denis revamped the grand spectacle tradition that first fueled her imagination. Popular commercial entertainments, such as Imre Kiralfy's *Nero, or the Destruction of Rome* (1888) and *Egypt Through the Centuries* (1892), encouraged a young St Denis to see the possibilities of refashioning a visual scope and theatrical framework for 'American' dance. The Lewisohn Stadium concerts offered her a mass forum and prestigous cultural vehicle for 'high art' to lead general audiences on a pilgrimage of hope during a period of personal, professional and social upheaval. She momentarily realized her quest to synthesize art and religion on a grand scale. The aesthetic pageantry found in this new period of work removed the exotic borrowings of non-Western myths and religions to communicate more immediately the exaltation of liberty. St Denis's esoteric design gave public expression to American ideals of progress, patriotism and a spirit of democracy embracing God and nation (Novak 1980; Glassberg 1990).

However, when St Denis returned from Asia, the landscape had changed dramatically. She was aware that a radical new dance movement was transforming the New York concert world by the late 1920s that would soon rebel against the established conventions of dance theatre Denishawn represented. Looking back over the last years in Denishawn House, St Denis wrote: 'I was fighting, sometimes jealously, sometimes despairingly, to retain something of my own confidence in the value and power of my destiny' (1939: 331). Her fate was caught in the transitions reshaping American culture. Julia L. Foulkes observes that 'by the end of the 1920s, the mission to create an American form of concert dance had begun in earnest, and the important role that social and political ideas would play in realizing this goal was set ... Artistic change meant social change as well' (2002: 25). In the field of concert dance the following decade erupted with revolutionary fervor.

One of the central debates within modern dance practices that arose in Depression-Era society contended over the cultural authority of dance (Graff 1997; Franko 2002). Independent 'solo' artists, like Martha Graham, appealed to artistic freedom (Soares 1992; Foulkes 2002; Siegal 1987). The formalism associated with this notion of aesthetic modernism attempted to cultivate an elite space for self-expression apart from politics. The left-wing radical dancers and collective dance groups such as the New Dance Group or the Workers' Dance League, who used the slogan 'Dance as a Weapon of the Revolutionary Class Struggle', pursued a political ideology that often supported socialist and Marxist communal structures. The collectivist ideal developed principles of mass ethos in both technical and aesthetic practices designed to represent the proletarian social body. Ellen Graff writes: 'the terms revolutionary and bourgeois most accurately describe ideological divisions existing between American dancers before 1934' (1997: 13). Both attitudes attacked the aesthetic framework of institutional and commercial choreographic practices while competing for mainstream acceptance. St Denis was percieved as part of the institutional and commerical terrain. Yet, the esoteric, Protestant lineage infusing her vision of nationalism reflects a third strand, one that signals a more persistantly porous tradition of popular idealism found in American theatrical spectaclism.

St Denis's reactions to the hotbed of aesthetic debates and divergent modes of choreographic production springing from the rise of an independent modern dance movement fueled a maturing artistic vision. She entered a new phase of work. Among these, *The Lamp* (1928) and *The Prophetess* (1931) were mystical allegories on the enlightenment of humanity. *Angkor Vat* (1930) attempted to resurrect the 'spirit' of ancient Cambodian civilization. What underscores the Lewisohn productions is an evangelical ferver that art elevates moral nature to higher spiritual truths, a liberation of the indivdual within divinly ordered, boundless, collective freedom. The androgynous nature of the universe is a creative impulse in St Denis's cosmology. It underlines her whole aesthetic philosophy. Knowledge of God involved the conjunction between aspects of the universe perceived as feminine - nature - and masculine - spirit. The union of male and female principles within a creative act is a generative process. St Denis believed the artist is constantly ascending to higher levels of spiritual understanding. It was this divine insight that culture communicated.

St Denis's sense of ritual was rooted in biblical themes and identification with non-cognitive essences. From an early age the rituals of the church were a 'irresistible attraction,' which she credits with developing her aesthetic sensibilities (St Denis 1939: 27). Biblical readings were not her only insight into spiritual awakening. St Denis was also exposed to popular literature from the many spiritualist movements sympathetic to non-Western religions, including Transcendentalism, Mesmerism, Theosophy and Buddhism, which were gaining widespread followings in the late nineteenth century. Works such as Edward Arnold's poem on the life of Buddha, *The Light of Asia* (1879), which became one of the most widely published accounts of Buddhism in the United States,[1] and Mabel Collins's esoteric saga *The Idyll of the White Lotus*, set in Ancient Egypt, depicting the spiritual initiation and persecution of a young man, incited her imagination with scenes of communion with divine and mystical transformations. Access to myths and stories from world religions opened her to a more expansive notion of 'cosmic consciousness' as a collective identity.

[1] Sales of *The Light of Asia* in the United States are estimated by Edward Arnold's biographer to be between 500,000 and 1 million copies (Tweed 1992: 29). St Denis heard a reading of *The Light of Asia* by Edmund Russell in 1905 (St Denis 1939: 59, 201-202; Shelton 1981: 49, 141).

Very often, discussions of Buddhism in America highlighted correspondences to Christianity: *The Light of Asia* emphasized parallels between the life of Jesus and Sakyamuni (Tweed 1992: 115-32). Ananda Coomaraswamy prefaced his introduction to Buddhist thought, *Buddha and the Gospel of Buddhism*, with ethical analogies between Christian mysticism and the Buddhist 'philosophy of Life'. Such literatures gave St Denis an outlet through which she could explore a less constricted image of herself than Victorian culture permitted. *Radical Spirits*, Ann Braude's 1989 study of the overlap between the Women's Rights Movement and nineteenth-century spiritualism points out how spiritualism offered such progressive women as evangelists, reformers and women's rights advocates greater individuality and economic opportunity. For one, spiritualism provided women with an ideological alternative to mainstream Protestant doctrines that held women subordinate to men. Traditional views on faith and religious practice came under revision. Spiritualism redefined one's relationship to God as a nonpersonal, metaphysical union. This independence granted women a forum for self-expression and degree of self-determination that upset marital and occupational norms.

St Denis's theological orientation was founded in one of the most influential spiritualist movements, Christian Science. Mary Baker Eddy published the foundational text of Christian Science, *Science and Health with a Key to the Scriptures*, in 1875; the same year Helena Blavatsky along with Henry Steele Olcott held the first meeting of the Theosophical Society in New York. In later years, St Denis became associated with this group. Theosophy melded assorted European and Eastern cosmic philosophies: Hindu Brahmanic meditation, mysticism, Buddhism, occultism and the esoteric theories of Emanuel Swedenborg (Swedenborg 1946 [1771]; Tweed 1992: 50-60). Theosophy's principles rest on divine intuition and proposed 'the regeneration of humanity and its reintegration into that state of primitive innocence preceding Adam's fall and original sin' (Knee 1994: 58-9). Divine knowledge was associated with noncognitive experiences. Intuitive processes were identified with a mysterious and mystical human nature capable of reaching higher levels of perception, or divine wisdom (Tweed 1992: 50-1). Blavatsky's two-volume treatise, *Isis*

Unveiled, published in 1877, advocated a return to the wisdom of 'ancient sages and philosophies,' which hold the 'secrets' to reunifying the individual in a cosmic order (Jackson 1981: 160). The re-examination of religious tradition, buoyed by occult and esoteric strains of spiritualism entering America from Europe and Asia in the last decades of the nineteenth century, underscore St Denis's own impulses towards reviving religious practices and sacred rites in dance.

Mary Baker Eddy also argued for an alternative approach to orthodox religious practice and asserted that Christian faith was a science of the mind and not a dogmatic system. She attempted to distinguish Christian Science from traditional Christian doctrine as a return to the 'objective' re-reading of Biblical scriptures. She was preoccupied with defining the nonpersonal, nonhuman nature of God, which made Christian Science popular among other schools of Spiritualism. 'Metaphysical or divine Science', Eddy states, 'reveals the Principle and method of perfection - how to attain a mind in harmony with God, in sympathy with all that is right and opposed to all that is wrong, and a body governed by this mind' (1886: 15). Traditional Christian teaching, she felt, offered mistaken views of Deity that evolved into a personal and materialistic view of God as human and a humanity directed by individual self-interest. Eddy's reading of scripture interpreted God's being as Truth, Life and Love, a 'trinity in unity; not three persons in one, but three statements of one principle' (1910: 3).[2] For Eddy, physical nature, including the body, blinded one from perceiving this spiritual law. Eddy repudiated the senses as false. Faith involved the power of mind over matter. In Eddy's metaphysics, ill health and mortality were signs of worldly entrapment, since the divine mind governs a perfect universe (9). Only by apprehending spiritual truth can one live in harmony with God. Communion with the divine inspires the conscious power to govern oneself in accordance with the revelatory motion of God's will. St Denis called Eddy's philosophy a 'science of consciousness' (1969: 58).

Eddy departs from traditional theological conceptions of revelation as an individual's redemption (Gottschalk 1973: 55). She maintains that God's presence is a 'self-revealing and sustaining power' (Eddy 1910 [1875]: 561). Revelation is the basic nature of divine activity - God's consciousness. Revelation is not a cognitive mediation between the human and divine realms; it is the complete presence of self in nature. 'In its mature stage,' writes historian Stuart Knee, 'Christian Science doesn't discuss heavenly ascent but spiritual selfhood realizable and attainable in the present' (1994: 59). Eddy's principle of revelation underscored an ethical system that placed the sign of moral virtue, of faith, in daily practice where one worked towards common ideals (Eddy 1910 [1875]: 5). The ethical component implied that self-awareness cannot be achieved without an embodied struggle. Over the years, St Denis's aesthetic concerns were tied to liberating the body from the constraints of time and space. Her conviction that the dance was part of artistic tradition, a vehicle of spiritual expression and a source of self-invention encouraged her to begin dreaming of raising a temple for dance that would have the 'motivations of the church with the instrumentation of the stage' (1939: 328).

'Dance,' St Denis claimed, 'must have its own temple'. Looking forward, St Denis and Shawn intitiated plans to build a training and performance complex, a 'Greater Denishawn'. They intended that this company would become the foundation of an American ballet. St Denis remained obsessed with trying to generate support for a theatre that would resemble an altar for dance. Neither the theatre nor the dream of Greater Denishawn was realized. Efforts to raise money independently to finance the project failed, St Denis's and Shawn's marriage ended in separation and the company collapsed. Not until the post-Denishawn years would St Denis make the church her stage. In the 1930s, she began a series of Madonna studies, notably her *Masque of Mary*, and dance liturgies that were staged in

[2] Swedenborg, defined the Trinity as 'soul, body and operation'. Operations are actions coordinated by body and soul. These essential aspects of God constitute the essence of being. Delsarte, whom St Denis studied, set forth in his Law of Trinity a synthesis of mind, body and soul. Following Swedenborg, Delsarte conceived the universe as consisting of three co-existent spheres: the natural, the human and the divine. The interplay of the material world, human consciousness and spiritual revelation flow back and forth, inwards and outwards, revealing 'correspondences' between the natural and the supernatural, the visible and the invisible, the material and the immaterial. The three realms of experience formed a unity of consciousness, mirroring the infinite motion of being (Swedenborg 1946 [1771]: 254).

church chancels (Shelton 1981: 242-3). St Denis was denied her temple to dance in. She converted the public space of Lewisohn Stadium into her cathedral of art.

3 The operating costs of the school at that time were $8,000. However, a large portion of the income earned from the concerts was characteristically absorbed in production costs. St Denis and Shawn earned a little over $8,000 for three performances in the summer of 1929. Half of that money covered production expenses (Shelton 1981: 230).

1. THE CONCERTS

Denishawn's celebrity was unparalleled in the United States when it began appearing at the Lewisohn Stadium Summer Concerts. The company was the first dance troupe to be invited to perform alongside the New York Philharmonic at the large outdoor amphitheatre situated on the grounds of City College in upper Manhattan. Twenty-thousand people 'stormed' Lewisohn Stadium for Denishawn's first appearance in July 1925 prior to the company's departure for Asia. The Doric-columned amphitheater was built in 1915, a gift from investment broker and philanthropist Adolph Lewisohn. The concert series was introduced in 1917 by a group of civic-minded New Yorkers, chaired by Minnie Guggenheimer. Mrs. Guggenheimer famously greeted the audience from the stage with an ebullient 'Hello Everyone', to which the crowd would shout back, 'Hello Minnie'. By the time Denishawn began appearing at the Stadium, the concerts were an established cultural venue and, as one critic wrote, 'an incalculable value to the New York public'. With an admission price that ranged from twenty-five cents to a dollar, the concert series was 'the best bargain in New York City', remarked one regular. While a large section of the audience was made up of serious music lovers and students, who arrived with 'score in hand', the vast majority had their first taste of orchestral music at the concerts. The programmes deliberately made 'no concession to public taste', but chose to offer a full range of symphonic music and opera. Other dance legends followed Denishawn. Michel Fokine presented his most best-known ballets, *Schéhérazade* and *Les Sylphides*, to record audiences in 1934 (New York Times).

The Lewisohn Stadium Summer Concerts remained the primary vehicle in which Denishawn regaled the public with extravagant new spectacles. They also brought in much-needed revenue to prop up the deteriorating financial situation of the institution.[3] Each production was staged with a view towards mesmerizing the large, cosmopolitan audiences. Typically, the 15,000 seat Stadium was filled to capacity for Denishawn's appearances. Preparing and rehearsing for the concerts occupied St Denis and Shawn year round. Scores of dancers - students, amateurs, and professionals - were assembled and shaped into 'Denishawn Dancers' (St Denis 1939: 332; Shelton 1981: 229-30; Mumaw and Sherman 1986: 27). The wholesome flair characteristic of Denishawn style aligned them with a cultured propriety approved of by a flourishing, white middle-class. St Denis's reorientation towards Western religious ritual channelled her notion of revelation into crowd-pleasing invocations of the mythic struggle to achieve enlightenment.

The Lamp

Denishawn returned to the stadium for the second time in August 1928 for three performances. The centrepiece of the programme was the premiere of *The Lamp*. The subject represented the transmigration of the soul - a theme she had attempted very early in her career in a dance based on Lafcadio Hearn's Japanese-influenced story *A Shirabyoshi* (1908). St Denis conceived an allegorical form to express the common principle that all enlightened beings were universally connected through knowledge of the divine. She expressed the Christian symbolism beneath a pageant of human discord and choric harmonies. The production was noted for its 'unmistakable mood and many moments of visual beauty'(1928).

The Lamp evolved out of a poem that St Denis composed on 'one of my nights spent alone in the big studio'(1939: 333). These were dark days when she was preoccupied with reconstituting her own life. She had visions of death and transfiguration. For the Lewisohn concert, she expanded this spiritual crisis into three movements: 'Life, Death and Transfiguration by Wisdom'.[4] The

4 The scenic descriptions for each production are compiled from St Denis's outlines and notebooks.

• **Jerome Robbins Dance Division,** *The New York Public Library for the Performing Arts, Astor, Lenox and Tilden Foundations.*

music was selected from Franz Liszt's symphonic poem *Les Preludes*. The choreography derived from St Denis's experiments with 'music visualization', a concept in which the dancers' movement embodied the musical structure. The costumes, by Pearl Wheeler, and languid compositional groupings were inspired from the drawings of William Blake. Wheeler's costumes were hybrid designs of pale gauze drapery, after ancient Greek sculpture, fastened in the style of Indian saris. They suggested a timeless vista of veiled archaic figures.

The only scenic element was an altar, a pyramid of steps placed centre stage and set off by a backdrop of black curtains. St Denis's role depicted 'the Woman with the Lamp', indicating enlightenment. Shawn played the figure of 'Life and Death'. The couple was joined by fifty dancers, who filled the huge stage, moving in mass patterns to the music or momentarily breaking up into separate tableaux. *The Lamp* symbolized 'perfect life', the eternal spirit, and was represented metaphorically on stage by a round lamp with many wicks carried by St Denis, who stood at the top of the steps. The underlying structure of the dance mirrored the four stanzas of St Denis's poem. Four 'rhythms' transpired in sync with the atmospheric transitions of Liszt's score. The last movement conveyed the coming of the Light as the ascent of the soul restores divine harmony to humanity.

The Lamp opens as the 'Divine Star of Life' (St Denis) appears at the top of the 'mountain'. The ensemble, with youthful abandon, dance joyously. But the figure of Death approaches, and the dancers freeze in terror. Shawn, dressed in black, wore a mask of death. This mortal figure spreads darkness and chaos, 'mowing down the youthful figures' as he sweeps in. He strides to the foot of the steps and strikes a triumphant pose. Then he directs a great veil - signifying burial - folded at the bottom of the steps to be unfurled over the bodies of the dead, who lie about the stage. At this point, St Denis rises from her throne of light and holds up her lamp. The floor of the top step was glass with lights beneath it, so that when St Denis stood up, she was bathed in brilliant blue light. The divine light attracts 'those who will come and light their torches' (Terry 1976:114-15). These figures prepare to become healers and seers, disciples who will spread wisdom throughout humanity. One effect created for the production was the use of the huge veil covering the floor of the stage beneath which dancers moved to make it billow. Shawn underwent his transformative rebirth - a change of costume - beneath it while the dancers manipulated the canopy. The use of the veil and washing the stage in blue lighting signified the transcendent moment.

The affirmation of placing art in the service of a collective-religious ethos found nourishment in St Denis's association with Claude Bragdon at this time. She met Bragdon through theatrical producer Walter Hampden around 1924 while involved with his production of *The Light of Asia* for the Krotona Theosophical Society in Hollywood, California. Architect, set designer, and publisher, Bragdon was an avid student of esoteric philosophy and occultism (1938). Active in the New York art world, he associated with photographer Alfred Stieglitz, scene designers Robert Edmond Jones and Norman Bel Geddes, and Sheldon Cheney, whose influential stewardship of *Theatre Arts Monthly* heralded the 'little theatre' movement and the 'New Stagecraft'.

Bragdon's own theories of art and architecture were influenced by his study of theosophy, transcendentalism and mysticism (1913; 1915; 1927). He became a prominent advocate of Fourth Dimension philosophy, which seeped into popular literature and modern art through mathematical and scientific investigations into non-Euclidean geometry. Bragdon's numerous writings on the subject contributed to its widespread popularity in the United States (Henderson 1983). The theoretical principles of Fourth Dimension philosophy, popularly known as hyperspace theory, generally proposed a non-linear space-time continuum (the n dimension). In 1920, he and Nicholas Bessaraboff produced the first English-language translation of Russian mathematician P. D. Ouspensky's mystical *Tertium Organum: A Key to the Enigmas of the World* (1912). Ouspensky's premise of a fourth dimension beyond known phenomena reinforced St Denis's pursuit of hidden realities. St Denis consumed Ouspensky's work, which, Suzanne Shelton points out, left its imprint on her thinking during the 1920s (1981: 177-80; St Denis 1939: 306-7).

St Denis had enlisted Bragdon with help in designing her own art theatre, which was never built.[5] His concept of organic architecture espoused a new 'space language' derived from Gothic architecture, which he argued released religious impulses in the nonphysical, rhythmic substance of everchanging form (Bragdon 1927: 123-44). Through the dance, St Denis reiterated Bragdon's view that form determined by the evolution of changing nonmaterial nature corresponds to psychic forces. The sensation of natural beauty is a sublimation towards 'the spirit of true democracy and true brotherhood' (142).

Angkor Vat

Angkor Vat marked Denishawn's fourth appearance at the Stadium. The music was composed by Sol Cohen, director of the Los Angeles Philharmonic Orchestra. The sets were designed by 'New Stagecraft' designer Cleon Throckmorton, who worked with the Washington Square Players and on Broadway with the Theatre Guild. The costumes were created by Wheeler. Shawn did not appear in the production. Instead, St Denis's partner was principal dancer Lester Shafer. The action of *Angkor Vat* represented 'a Khmerian day'. In scope, the production attempted to rekindle the golden age of Cambodian history (AD 802-1431), when the temples and cities of Angkor were built. More significantly, at the advent of the Khmer Empire, Jukka Meittinen remarks, 'Khmer dance achieved the status of a kind of state art' (1992: 141). In Angkor, St Denis discovered a supreme example of a cultural ideology where art and nature intermingled in the social consciousness.

Angkor Vat evolved out of a visit to that ancient city in 1926. Towards the end of the Asia tour, the company appeared in Saigon. St Denis, Shawn, Weidman and Humphrey took a side trip to Phnom Penh to attend a birthday celebration for King Sisowath, where they saw a performance by the Royal Cambodian Ballet and visited Angkor. St Denis had investigated aspects of Khmer court dance in the past with her *Siamese Ballet* (1918). A *Cambodian Ballet*, attempting to imitate the Royal dancers, was slipped into the Denishawn repertory on its return from Asia (Humphrey 1972: 59-60). In her notes for *Angkor Vat*, St Denis wanted to express the sense of awe she felt while standing amid the ruins of that highly advanced civilization. *Angkor Vat* took shape from the sculptures and carvings of dancers she

[5] The design for the theatre never appears to have developed beyond Bragdon's sketches and a preliminary drawing by architect Hugh Ferris. Among St Denis's notebooks is a sketch for a temple set that details a spherical stage ringed by mirrors, panels and a circular lighting track. It is dated 26 December 1927.

saw on the temple walls (Meittinen 1992: 141). St Denis used these visual sources to reconstitute a historical perspective in which dance played a central role in society.

St Denis's reveries supplied the narrative. She 'imagined an artist, sitting in the scattered shade of a palm tree with the silence of the jungle about him and the great dead city in front of him'. The artist is in the process of transferring the movement of the sculptured figures on the temple to canvas. He stops painting, and, leaning against a tree trunk, begins to daydream about the life of this civilization. The figures on the temple walls slowly begin to emerge from their immobility. 'He see them take on a stylized sort of life ... after the manner of ... those incredible artists who gave them an immortal birth in stone'. The set was painted with sand and earth colours to resemble a jungle. The dancers wore costumes of the same shades, their bodies tinted with brown paint. The production had a company of forty dancers, many of whom were amateurs. Barton Mumaw was part of the company - in fact, it was his first appearance with Denishawn - but he twisted his ankle and could not perform. Mumaw recalled watching the performance from the stadium: 'Every seat begin to fill as the sky darkened. The lights above me dimmed. The huge audience fell silent after the conductor took his place before the New York Philharmonic Orchestra'(Mumaw and Sherman 1986: 28-9). Beneath the elaborate lighting, each intricate,

• **Jerome Robbins Dance Division,** *The New York Public Library for the Performing Arts, Astor, Lenox and Tilden Foundations.*

crumbling sculpture came to life.

The first scene depicted a watchman opening the great gates of the king's palace, the milkman bringing milk, a street sweeper and girls getting water. The cast, which included Jack Cole, Florence Lessing, Anna Austin and Klarna Pinska, took up poses resembling the ancient friezes. The daily activities unfolded into preparations for a feast given by the king and entertainment by dancing girls, who moved in unified patterns after the dancers St Denis had seen carved on the temple walls. The festivities were interrupted by a battle, enacted with the dancers symbolically riding the backs of elephants, which they suggested with their own bodies.

The climax was a love duet between St Denis and Shafer. St Denis portrayed the Naga Queen. The Naga, or sacred cobra, in Khmer culture is believed to be the progenitor and protector of the royal line. It is often depicted with seven heads. St Denis drew from the many images of the Naga to create one of her most extraordinary personifications. Advance press for the Stadium concerts featured her writhing in the form-fitting serpentine costume, trailing a six foot long tail that Wheeler designed for her. Her headdress replicated the towering cobra-hood, studded with seven cobra-heads, found guarding the city gates and entrances to the temples in Angkor.

There is a popular legend associated with the great serpent, which St Denis adopted. It was alleged that in order to ensure the continuity of the kingdom, every night the monarch coupled with the sacred serpent, which manifested itself in the form of a woman for this nightly ritual (Myrdal and Kessle 1970: 81-7). The final scene concluded with the nightly nuptials of the king and the Naga. To St Denis, the Naga was an androgynous deity, a source of daily regeneration that personified 'cosmic energy'. A final tableaux expressed the completion of the union as St Denis and Shafer entwined themselves in a mirror-image of one another.

St Denis's Naga Queen exposed a more aggressive side of the feminine than any of her prior goddesses. In her first appearance as the great serpent, St Denis danced with masculine force. She repeatedly raised one arm over head, then struck her chest to suggest animal vitality and male strength (Shelton 1981: 233-4). The Naga signified that power and beauty reside in nature. St Denis believed that when the animal and masculine aspects of an individual are in concert through devotion to God, spiritual ethos is restored. In Angkor Vat St Denis recast herself as a civilizing spirit. In her next production, St Denis arrived at the Stadium portraying a savior of the world.

The Prophetess

A summer of extreme heat and financial depression did not affect the enormous attendance at the Lewisohn concerts in 1931 (*New York Times*). Denishawn's fifth season at the Stadium drew standing-room crowds. On its second night extra seats were installed to accommodate the 'throngs' of fans that unknowingly watched Denishawn's last public performance. St Denis's *The Prophetess*, along with Shawn's *Job, A Masque for Dancing*, were the featured new spectacles. The dance's inspirational message, adopted from Christian Science 'healing', clearly reflects St Denis's response to the deepening depression across the country. The figure of the Prophetess (St Denis) symbolized all enlightened women. At first she appears the dark prophetess, the hope of the world, but she is crushed by it. Knowing that the indivisible structure of the world rests upon balance, rhythm and beauty, the Prophetess moves on the mountain top in a divine dance and is mirrored by humanity below.

St Denis led a company of fifty-five dancers, who peopled the stage with contemporary workers, artists, sailors, farmers, housewives, school children and 'other representatives of the mass of humanity'. St Denis wore a red, long-sleeved gown with a front zipper. She slipped it off at the end, revealing underneath priestly white robes. The main scenic element was the pyramid of steps used in *The Lamp*, which was also included on the programme. The three scenes depicted the phases of the soul's rejuvenation: compassion, the struggle to release oneself from material and mortal bonds, and enlightenment. They corresponded to the dissonant 'Mars' movement of Gustav Holst's *The Planets*. The last movement resounded with harmony: 'One after the other they tune in with her rhythm until she leavens the whole mass to the joy of rhythm and of spirit'. The finale was danced and sung to a hymn by English clergyman John Bacchus Dyke, 'Holy, Holy, Holy'. From the stage, the company encouraged the audience to sing along.

The Prophetess, according to St Denis's manuscript notes, represented the 'Woman Thinker'. St Denis saw the role of the prophet as one who leads a people to a vision of spiritual unity, and that of the artist as one who reveals the beauty and harmony of such a spiritual life. She identified herself with 'all women of illumined vision', citing Miriam, sister of Moses, Joan of Arc and Mary Baker Eddy among those who divined a path and exulted in God's work. She drew on several sources that underscored the visionary panorama of urgent despair and human frailty hovering at the feet of an eternal mother. The philosopher Manly P. Hall - a personal friend of St Denis's who founded the Philosophical Research Society in 1934 - reinforced for her the idea that greed and selfishness were destroying modern society.[6] Albert Einstein's thought supplied a definition for cosmic religion, the highest stage of religious belief without dogmas. This condition is achieved when individual destiny is seen as an imprisonment and a collective existence is formed. From biologist Albert Edward Wiggam's essay *The New Decalogue of Science* (1923), St Denis derived the notion that society must see the arts as an ark carrying it forward (St Denis 1939: 329).[7]

St Denis also returned to her beloved Mary Baker Eddy. No longer content to dress the iconography of divinity in ancient myths and Asian goddesses, St Denis portrayed an agent of social renewal. In Eddy's account of the 'Book of Revelation', she likens woman to the 'spiritual idea of God' (Eddy 1910 [1875]: 561). Woman is 'spiritual sunlight' and the spiritual idea is 'typified by a woman in travail, waiting to be delivered of her sweet promise' (562). In this sense, the body of woman brings forth light. The complete spiritual being, in St Denis's view, manifested the essence of God's motherhood - the androgynous energy of the soul - as the catylist for social revitalization.

In St Denis's earliest portraits of feminine icons, *Radha* and *Egypta*, she focused on self-liberation. *Radha* embodied freedom in the release of the body from its physical bonds; *Egypta* represented the enlightened woman. Informing St Denis's thought at this time was Eddy's proclamation: 'I saw before me the sick, the lame and the dying - slaves to the unreal Master, and I wanted to save them from the bondage of their own beliefs.' *The Prophetess* envisioned St Denis's ultimate aspirations: the 'dance of confidence and joy, where inspiration not for the few but for the many is attained' (1939: 328). The true act of freedom, of worship, was to liberate the social body from the shackles of materialism and the threat of totalitarianism by reviving American ideals through 'eternal truths'.

2. TRANSITIONS: INDEPENDENT MOVEMENT AND THE SOCIAL BODY

The Lewisohn productions were evangelical statements that pit St Denis against the current secular 'freedoms' blossoming across the spectrum of dance styles in the late 1920s. St Denis viewed the shift towards new experimental forms of dance as a threat to her preeminence. She answered by upholding the *lamp* of tradition.[8] In fighting for her artistic legacy, she clung to what had become a conservative agenda that freedom - as a source of cultural enlightenment - was part of Americans' spiritual identity. Her belief system was in complete opposition to the polemics of period. She saw her values being torn asunder by personal aesthetics and artistic priviledge and even more by political contention over the destiny of America (Kennedy 1999). In 1928,

[6] Quoting from Hall's *Lectures of Ancient Philosophy: An Introduction to the Study and Application of Rational Procedure* (1929), St Denis pursued his aesthetic theory: 'When we love the beautiful as we now love the dollar, we shall have a great and enduring civilization.' Hall's interests concentrated on Eastern and esoteric philosophies, occult studies and astrology.

[7] Wigman's essay elaborated the scientific principles supporting the neo-Darwinist theory of eugenics. The *New Decalogue of Science* first appeared in the *Century Magazine* (March 1922). St Denis repeatedly quotes Wiggam's declaration: 'Art is the Ark of the Covenant in which all ideals of beauty and excellence are carried before the race.'

[8] In 1913, the introduction of modern European art at Armory Show in New York launched virilent attacks from the American academic painters. Among them Kenyon Cox responded with a painting of a maidenly figure symbolizing tradition carrying an oil lamp lit by the everlasting torch of the beautiful (Shapiro 1978: 165).

St Denis and Shawn wanted to instil a a quota system in the Densiahwn school intended to limit Jewish students 'to ten percent of the whole' (St Denis 1939: 344; Humphrey 1972: 64; Graff 1997: 20). Fifty-eight people immediately left the school (McDonagh 1973a: 19). Russian Jews were principal targets of anti-Communist rhetoric and immigrations policies in the 1920s (Gerstle 2001: 81-127). Such a racially targeted quota system may have stemmed from her motivation at this time to reaffirm a Christian heritage through dance performance. St Denis eventually became involved with the Oxford Group, an elitist and conservative evangelical Protestant association that promoted anti-communism and pacifism. The group was led in the United States by St Denis's friend, the Reverend Samuel M. Shoemaker, rector of Calvary Episcopal Church in New York (St Denis 1939: 384; Shelton 1981: 246-7).

Since Martha Graham's own 'defection' from Denishawn in 1923, she had reunited with Louis Horst, the company's former musical director, produced her first concert, which set the New York dance world buzzing, and opened her own studio (Soares 1992; McDonagh 1973b: 50-2, 303-5). By 1927, Graham began to find her own stark movement vocabulary. The premiere of *Revolt* confounded public perceptions of aesthetic dance (Franko 2002: 158-62; McDonagh 1973b: 58; Graff 1997: 59). This future *grande dame* of American modern dance, along with her contemporaries and European counterparts, Mary Wigman, Harald Kreutzberg, Yvonne Georgi and Hanya Holm, joined the 'new' dance movement in New York that hatched a tumultuous era of modern experimentation. Don McDonagh alludes to the rising tide of post-Denishawn dancers as the 'generation of unknowns' (1973b: 58).

Doris Humphrey, Charles Weidman and Pauline Lawrence named themselves 'the unholy three'(Humprhey 1972: 82). This close-knit trio was the first within the company to revolt against St Denis's and Shawn's vision of Greater Denishawn (Siegal 1987: 72-7). Humphrey knew she was at a crossroads in her career by the time she returned home from Asia in 1926. Still under the spell of that extraordinary tour, she contemplated her own future as a choreographer. After ten years with Denishawn, Humphrey was aching to break away but held back out of loyalty and the inability to leave the security of her position within the company. A chance meeting with Martha Graham over the summer of 1927, in which the two future dynamos of American dance shared a friendly exchange, reinforced Humphrey's desire to strike out on her own.

Humphrey was almost thirty-three. Lawrence was twenty-eight, and Weidman, the youngest of the group, was twenty-seven. They were mature professionals, bursting with talent and of a generation that included Anna Sokolow, Jane Dudley, Katherine Dunham, Jack Cole and Agnes de Mille, who were poised to seek their own form of expression through dance. Weidman and Humphrey were no longer content to play ethnic, exotic or character roles. They challenged the essentialist aesthetics of St Denis's hybrid mythology, seeking instead a source of self-expression that was independent of cross-cultural borrowings (Partsch-Bergsohn 1994: 63). Weidman, Walter Terry relates, 'began to wonder, "What am I?"' (1971: 116). Humphrey wanted to experiment with movement that could speak about 'ourselves as Americans' (1972: 63).

Graham's debut as an independent artist, and Humphrey's, Weidman's and Lawrence's departure from the company were merely one indication that the currents in concert dance were turning against Denishawn tradition. The field of dance, in recitals, in theatres, from folk forms and ethnic styles to ballet and social dance, garnered attention with seventy-nine performances in the first few months of the 1928-9 season. Mary Wigman made her first concert tour in the United States in 1930, beginning with eleven performances at the Chanin 46th Street Theatre (Odom 1980; Manning 1993). The start of her ten-week tour was surrounded by enormous publicity. The sold-out house overflowed with professional dancers, modern dance enthusiasts and critics on opening night. At the end of the concert, the audience went wild and refused to leave the theatre until Wigman gave an encore.

Wigman's appearances followed on that of her students, Kreutzberg and Georgi. These concerts were part of a broader infusion of international artists that awakened the American avant-garde to its own potential. The famed Spanish dancer La Argentina, Peking Opera star Mei Lanfang and Indian poet and playwright Rabidranath Tagore, who collaborated in a dance and poetry recital with St Denis, all toured the United States in 1930. Less than a month after Wigman's concerts Humphrey, Weidman and Lawrence - now the Humphrey-Weidman troupe - teamed up with Graham and Helen Tamiris, who turned to modern dance from ballet, to launch the Dance Repertory Theatre. The first season took place over three nights in January 1930 at the Maxine Elliot Theatre in New York. In attendance was a young dance critic named John Martin, who was already championing the principles of the 'new' dance in the pages of the *New York Times* (1929).

Martin was one of the first full-time dance critics hired at a major New York newspaper. The *New York Herald Tribune* hired Mary Watkins in 1927. Shortly after, Martin joined the *New York Times*.[9] These positions were created because of intensive lobbying from a ballooning urbane dance community. St Denis and Shawn took a vigorous part in the campaign. Martin evolved into the most influential voice of the German-American aesthetic. German dance critics of the 1920s coined the term 'absolute dance' to describe the new dance movement, know as *ausdruckstanz* (dance of expression). Martin translated the concept as 'expressionist' dance. He emerged at the head of a critical establishment that by the 1930s was in process of defining a theoretical and historical account of American dance modernism as aesthetic independence. Critics like Martin, advocating modern dance, and his counterpart in the field of American ballet, Lincoln Kirstein, helped promote, validate and institutionalize the American equivalent to European high art forms. Historical narratives that appeared in the 1930s, such as Martin's *The Modern Dance* (1933) and Lincoln Kirstein's *The Book of the Dance* and *A Short History of Classical Theatrical Dancing* (1935) laid the foundation from which a critical tradition pursued definitive notions of high art in terms of the chronology of American ballet and dance formalism.[10] Lincoln Kirstein is one of the key figures in the development of American ballet. With Edward M. M. Warburg he established the first school of American Ballet under the direction of George Balanchine in 1933. From the school, the American Ballet Company was formed, which became the resident company of the Metropolitan Opera in 1934. St Denis attempted to recover her own contributions to American dance history with the publication of her autobiography *An Unfinished Life* in 1939.

Influenced by Wigman's example, Martin expounded on the theoretical framework and technical innovations of absolute dance (1965 [1933]: 13-16). He advocated the necessity of establishing new standards and codified training systems for the new dance forms energizing the American concert stage. Martin acknowledged St Denis and Shawn as pioneers of American dance. However, he categorized St Denis's contributions, along with those of Isadora Duncan, as part of the romantic tradition, largely because he considered interpretive dance technically an uncodified choreographic system (Martin 1965 [1933]: 1-33). Moreover, the pictorialism, narrative structure and theatricality characteristic of Denishawn style reflected components of commercial - lowbrow - spectacle. It is likely that Martin is the unnamed reviewer who considered a large portion of Denishawn's 1930 Lewisohn Stadium concert 'near the beaten track' and *Angkor Vat* 'long, pretentious and dull'(1930). Martin maintained that St Denis was an offshoot of theatre culture. Her depictions of 'Oriental' subjects typified her influence on stage dancing as a 'character dancer' (Martin 1968 [1936]: 132). St Denis believed that Martin encouraged a rebellion

9 Walter Terry replaced Watkins at the *New York Herald Tribune* in 1939 (Terry 1976: 124-5).

10 *The Modern Dance* was based on series of lectures-demonstrations Martin gave at the New School in 1931-2.

against the system of Denishawn, particularly among her former dancers, Graham, Humphrey and Weidman, whom he lauded. She viewed the advance of expressionist dance with ardent distaste (St Denis 1939: 325). Echoing Bragdon, St Denis argued that absolute dance was 'intellectual rather than spiritual' (Bragdon 1927: 128). 'There is no such thing as absolute dance,' she remarked in one essay, 'for we are not yet disembodied spirits.'

St Denis focused her objections on two issues: one was the personal rather than impersonal search for artistic expression; the other was the emphasis on physicality. She could not understand the formalistic properties of dance that embodied movement for its own sake. It was overly athletic and, in her opinion, was taught 'from the mind and from the body, never from the heart or soul'. She was at odds with the dissonant and reflexive individualism manifested in the dancer's body. It was unmediated, raw subjectivity; it was not beautiful. The exploration of subjective impulses for its own sake, which priviledged form over content did not aspire to higher moral principles or extend beyond individuality. In St Denis's view, the dance was a medium of worship in which subjectivity transported the individual into a sublimated union with the rhythm of divinity. St Denis aspired to convey an aestheticized consciousness. The Beautiful was the presence of being in perfect harmony with time, when the invisible nature of spirit was embodied. St Denis was repelled by the idea of a body in disharmony with itself, because that signified an imbalance of nature and discontinuity with the universe. Indirectly, St Denis related the formal values of absolute dance to cultural decadence. She objected to the accent on the pelvis in the technical vocabulary of the 'German method'. The distorted, jagged thrusts and lunges emanating from the torso suggested to her only masculine, aggressive and mechanistic qualities. She had, after all, spent her career trying to release the body from its material shell and induce a dematerialized realm of consciousness. When confronted with a bodily identity struggling against itself, the sheer physical presence appeared hard, overemotional and grotesque.

When the young dancers in Denishawn expressed the view that the dancer should respond to current times, St Denis countered: 'Fear, hate or degeneracy can never in themselves produce an enduring art' (1939: 327). Carrying her argument further, on reflection, she related:

> Yes, I would have said, we are living in intense and turbulent times where every last ounce of human wisdom is needed. But I want my beloved art to contribute to the hope and exaltation of humanity by its gifts of rhythm and beauty, and the glory of a final human victory.(327)

St Denis convinced herself that the interest in the new dance was a phase. She redirected much of her energy into overseeing her Society of Spiritual Arts (later the Church of Divine Dance) and producing religious pageants and dance liturgies. She continued composing temple ballets centred on Christian themes and private worship. With her rhythmic choir, she performed these biblical pageants in Church sanctuaries. The first, titled Babylon, represented the salvation of a wayward woman. It was presented at Calvary Episcopal Church in New York in 1937.

St Denis's position within the turmoil of aesthetic differences opposed both the abstract formalism of modern dance and the politics of social activism. The Christian values she expounded were antithetical to the international sophistication of the new intelligentsia and new audiences that supported concert dance. With the proliferation of dance genres and striated modes of cultural production, her former stature shrank. Her metaphysical ideals belonged to a developing culture industry from a pre-World War I generation in which high art forms and popular culture intermingled to became democratizing entertainments. Aesthetic modernism and radicalism came to the forefront of dance, and with that a critical tradition, an institutional apparatus and patronage emerged that devalued the popular and commercial roots of American dance. St Denis's connections to vaudeville and commercial entertainment interfered with her attempt to cultivate a pedigree as serious art within contemporary artistic debates. She intimated her concern that 'there even grew up a rumor that instead of giving our art in a pure form we were popularizing it' (1939: 306). Indeed, by the 1930s, the popular saying in the entertainment industry was 'scratch a dancer and you find Denishawn' (Sherman 1979: 2). Marcia Siegal points out 'almost every modern dance choreographer can be traced back, via at least one chain of development, to Denishawn' (1979: 11). The effect of Denishawn's multilayered positioning in progressive education, in the entertainment field and in the modernization of dance production enabled its students and teachers to generate 'a new relation between dance performance and American culture' (Thomas 1995: 165).

St Denis's strongest objections reveal her intolerance for what she felt was the 'materialistic atmosphere' of modern post-World War I society. She believed that the cultural values expressed in modern dance were being motivated by secular concerns, while religious dancing issues from spirituality (St Denis 1997: 33-8). These two forms supported oppositional ways of life in her view. It was a sign of cultural 'negation, impermanence and death' (1939: 327). Society was in moral jeopardy. Her answer sought the recovery of a cohesive structure in which the artist would light the way to a better world: 'a world where we can live and move and have our being to the full - where we can radiate in all directions of time and space the things that are in our hearts'. Her concept of dance drama liberated the flesh to express interactions of feeling and sublimation, sense and spirit, illusion and enlightenment, which offered an evocative, beautified image of freedom. Her spiritual quest was to redefine the individual's relationship to society; dance was her conduit of social communion. St Denis's rebuttal to the divergent and divisive strains of Depression-era life was to affirm her idealism and perpetuate a Christian agenda. The Lewisohn Stadium concerts served as a pulpit to preach her ritualized liturgies to the masses. In the preface to *The Prophetess St Denis* proclaimed, 'Be with us, in this new rhythm of the dance of life.'

REFERENCES

Bragdon, Claude (1938) *More Lives Than One*, New York: Alfred A. Knopf.

Bragdon, Claude (1927) *Six Lectures on Architecture.* Chicago: University of Chicago Press.

Bragdon, Claude (1915) *A Primer of Higher Space (the Fourth Dimension)*, to which is added 'Man the Square, A Higher Space Parable', Rochester: Manas.

Bragdon, Claude (1913) *Projective Ornament*, Rochester: Manas.

Braude, Ann (1989) *Radical Spirits: Spiritualism and Women's Rights in Nineteenth-Century America*, Boston: Beacon Press.

Eddy, Mary Baker (1886) *Christian Healing and Other Writings*. Boston: Christian Scientist Publishing.

Eddy, Mary Baker (1910) *Science and Health With Key to the Scriptures*, Boston: Christian Scientist Publishing.

Franko, Mark (2002) *The Work of Dance: Labor, Movement and Identity in the 1930s*, Middletown, Connecticut: Wesleyan University Press.

Foulkes, Julia L. (2002) *Modern Bodies: Dance and American Modernism from Martha Graham to Alvin Ailey*, Chapell Hill: University of North Carolina Press.

Gerstle, Gary (2001) *American Crucible: Race and Nation in the Twentieth Century*, Princeton: Princeton University Press.

Glassberg, David (1990) *American Historical Pageantry: The Uses of Tradition in the Early Twentieth Century*, Chapel Hill: University of North Carolina Press.

Gottschalk, Stephen (1973) *The Emergence of Christian Science in American Religious Life*, Los Angeles: University of California Press.

Graff, Ellen (1997) *Stepping Left: Dance and Politics in New York City*, 1928-1942, Durham: Duke University Press.

Henderson, Linda Dalrymple (1983) *The Fourth Dimension and Non-Euclidean Geometry in Modern Art*, Princeton: Princeton University Press.

Humphrey, Doris (1972) *Doris Humphrey: An Artist First*, ed. Selma Jeanne Cohen, Middletown, Connecticut: Wesleyan University Press.

Jackson, Carl T. (1981) *Oriental Religions and American Thought: Nineteenth-Century Explorations*, Westport, Connecticut: Greenwood Press.

Kennedy, David M. (1999) *Freedom From Fear: The American People in Depression and War, 1929-1945*, Oxford: Oxford University Press.

Knee, Stuart E. (1994) *Christian Science in the Age of Mary Baker Eddy*, Westport, Connecticut: Greenwood Press.

Manning, Susan (1993) *Ecstasy and the Demon: Feminism and Nationalism in the Dances of Mary Wigman*, Berkeley: University of California Press.

Martin, John (1965 [1933]) *The Modern Dance*, New York: Dance Horizons.

Martin, John (1968 [1936]) *America Dancing*, New York: Dance Horizons.

McDonagh, Don (1973a) 'Conversation with Gertrude Shurr', *Ballet Review* 4.5: 19.

McDonagh, Don (1973b) *Martha Graham: A Biography*, New York: Praeger.

Meittinen, Jukka O. (1992) *Classical Dance and Theatre in South-East Asia*, Oxford: Oxford University Press.

Mumaw, Barton and Sherman, Jane (1986) *Barton Mumaw, Dancer: From Denishawn to Jacob's Pillow and Beyond*, New York: Dance Horizons.

Myrdal, Jan and Kessle, Gun (1970) *Angkor: An Essay on Art and Imperialism*, trans. Paul Britten Austin, New York: Pantheon.

Novak, Barbara (1980) *Nature and Culture: American Landscape Painting, 1825-1875*, New York: Oxford University Press.

Odom, Maggies (1980) 'Mary Wigman: The Early Years, 1913-1925', *The Drama Review* 24.4 (December): 81-92.

Partsch-Bergsohn, Isa (1994) *Modern Dance in Germany and the United States: Crosscurrents and Influences*, Switzerland: Harwood Academic Publishers.

St Denis, Ruth (1939) *An Unfinished Life*, New York: Dance Horizons.

St Denis, Ruth (1969) *Reflections on a Life in the Dance*, Oral History, interviewed by Elizabeth I. Dixon, Ruth St Denis Collection, University of California at Los Angeles.

St Denis, Ruth (1997) *Wisdom Comes Dancing: Selected Writings of Ruth St Denis on Dance, Spirituality and the Body*, ed. Kamae A. Miller, Seattle: Peace Worker.

Siegel, Marcia B. (1987) *Days On Earth: The Dance of Doris Humphrey*, New Haven: Yale University Press.

Siegel, Marcia B. (1979) *The Shapes of Change: Images of American Dance*, Boston: Houghton Mifflin.

Shapiro, Meyer (1978) 'Introduction of Modern Art in America: The Armory Show', *Modern Art, Nineteenth and Twentieth Centuries: Selected Papers*, New York: George Braziller: 135-78.

Shelton, Suzanne (1981) *Ruth St Denis: A Biography of the Divine Dancer*, Austin: University of Texas Press.

Sherman, Jane (1979) *The Drama of Denishawn Dance*, Middletown: Connecticut: Wesleyan University Press.

Soares, Janet Mansfield (1992) *Louis Horst: Muscian in a Dancer's World*, Durham: Duke University Press.

Swedenborg, Emanuel (1946 [1771]) *The True Christian Religion Containing the Universal Theology of the New Church*, vol. 1, trans. John C. Ager, New York: Swedenborg Foundation.

Terry, Walter (1971 [1956]) *The Dance in America*, New York: Harper and Row.

Terry, Walter (1976) *Ted Shawn: Father of American Dance*, New York: Dial Press.

Thomas, Helen (1995) *Dance, Modernity and Culture*, London: Routledge.

Tweed, Thomas A. (1992) *The American Encounter with Buddhism, 1844-1912: Victorian Culture and the Limits of Dissent*, Chapel Hill: University of North Carolina Press.

Walking, Standing, Sitting Like a Duck
Three instances of invasive, reparative behaviour

CAROL BECKER

> The British critic George Steiner claims that art, like certain kinds of religious and metaphysical experience, is the more 'ingressive', transformative summons available to human experiencing ... It is an intrusive, invasive indiscretion that 'queries the last privacies of our existence'; 'an annunciation that breaks into the small house of our cautionary being', so that 'it is no longer habitable in quite the same way as it was before'. It is a transcendent encounter that tells us, in effect: 'change your life'.
>
> Karen Armstrong (2005: 148)

In the traditions of monks and renunciates of both the East and the West, there are those who settle into a monastic community and spend a lifetime cloistered in one place pursuing a spiritual path. And there are those who are nomadic, moving from monastery to monastery, in a pattern that can take a decade to complete, stopping at each location for a circumscribed time - the journey, the pilgrimage, the ellipse of departure and return the defining principle of their lives. The desert fathers and mothers who lived in Egypt in the fourth and fifth centuries C.E. in the Roman Empire were a group who made a point of not settling anywhere for too long, of meditating continuously, of never taking comfort in a real home, family or a traditional, stable community. They were attempting to achieve *amerimnia* (freedom from care) and *apatheia* (detachment) (Caner 2002: 15, 92-3). They hoped 'to seek out the desert frontiers' in order to 'become strangers to the world' *(xeniteia) (2002: 25)*. By the fifth century there were reports of monks populating the desert. John, an elder known for 'surpassing all other monks in virtue', had stood motionless for three years under a rock in prayer (30-31). But when his feet gave out, an angel appeared, healed him and ordered him to move on. 'Thereafter, we are told, 'he lived wandering in the desert, eating wild plants' (31).

Although in practice such monks always needed to stay close to the towns for access to water and supplies, they essentially had moved from the comfort of community to the expanse of unpopulated wilderness. In the actions of such monks we can observe the uniqueness of lives dedicated to the spiritual, in direct contrast to those of the 'mundane shell', as William Blake might call it.[1] What function did these monks and their individual struggles serve for the adjacent communities? In their own terms, they were hoping to achieve a greater level of devotion and to find God in isolation, as Christ had done in the desert. They saw themselves as sinners - poor, weak men - but others saw them as '[t]rees ... purifying the atmosphere by their presence' (Russell 1980: 12).

In their refusal to work at any task other than that of achieving this type of spiritual elevation, these ascetics, not all of whom were monks, generated both adoration and disgust. Some people saw them as representing an ideal of purity in their attempts to make themselves Other to the world. But there were those who regarded the monks' refusal to work while begging for food as indicative of nothing more than sloth.

What is the meaning of those who choose to

[1] In Northrop Frye's definitive text, *Fearful Symmetry: A Study of William Blake*, he discusses Blake's use of this term in relationship to the prophetic books like *Jerusalem*. Frye writes, "In Blake the Firmament is the Mundane Shell, the indefinite circumference of the physical world through which the mind crashes on its winged ascent to reality."

Performance Research 13(3), pp.139-145 © Taylor & Francis Ltd 2008
DOI: 10.1080/13528160902819455

live an unearthly life on earth, a life dedicated to the task of turning the inside out - performing the public act of private contemplation? How is the circle of the ordinary broken so the extraordinary can be let in? We know that humans often live within multiple states of consciousness simultaneously, although they appear to be moving through space with the same intentions, with family and work supposedly at the core. Yet in each person there is an imaginary, intuitive, creative self filled with aspirations that usually remain hidden. Only when humans make small or large gestures within the public arena - personal acts of protest; public acts of protest, such as marches, sit-ins, strikes and social actions; or art designed to shatter, to rupture the continuity - that we become aware to what degree the status quo for many has become a pressure cooker exuding steam.

Right before the decision to invade Iraq was made public in 2003, there were massive protests in the US and elsewhere. Millions of people in the U.S., at least, led mostly by high school and college students, took to the streets in an exhilarated manifestation of disapproval, reminiscent of the student movements of the Vietnam War years and the Free Speech and Civil Rights Movements of the 1950s and 1960s. These actions of 'taking the streets', 'taking back the streets', putting a stop to business as usual represent a type of urban domestic warfare against the power structures that recalls the notion of participatory democracy as theatre. These ordinary/extraordinary gestures of individuals generate collective excitement, allowing people to feel that the world could be renewed and reinvented and that they could be part of its transformation.

In Hannah Arendt's sense, true *action* is an assertion of palpable, spontaneous, grassroots power. She also writes, .action is risk' (Young-Bruehl 2006: 89). According to Elisabeth Young-Bruehl, Hannah Arendt held that the pre-platonic Greeks, the original theorists of action, 'understood that action was dependent not on organized or legislatively created spaces for action, or on political organization or governments', but 'simply on people coming together to share words and deeds - "the unreliable and only temporary agreement of many wills and intentions" ... This coming together of actors is what Arendt called power' (89). To Arendt this power could be individual, shared and collective, or it could find its locus in a place where individual action and collective action work together as a common force. She wrote, 'action is judged for neither its motivation nor its aim, only its performance' (87).

So when one participates in this type of action, the sheer unexpectedness of it, the outrageousness of it, the risk of it make one feel, on one hand, that it seems highly improbable that such actions manifest at all, that people *ever* come together with such bravado. On the other hand, one also recognizes that it is surprising that such ruptures do not occur more often, that people do not congregate regularly to change existing conditions or that they do not assert more lone manifestations of social action. If we are unhappy with the status quo, such actions are exhilarating. If we are fearful of change, they are a source of anxiety, hence the repressive degree to which governments quickly constrain such actions to reestablish order when it appears threatened.

In America, a notoriously utilitarian society, art can function as a type of intervention in that it makes apparent the ephemerality and individuality of consciousness. Art that is spontaneous, risky and placed in the public arena - not as permanent sculpture but as an organic response to particular historical, physical or environmental conditions - is particularly unexpected in a market-driven economy and is therefore potentially transgressive. And even more rare is artwork that attempts to transform the solitary gesture of spiritual or political regeneration into a public performance. How does one come to make such work? How does one access its results?

• *(left)* **Ernesto Pujol, Walk #1 performance, Charleston,** *Digital Image, 2004-2005*

• *(right)* **Ernesto Pujol, Mourning Circle performance, McNay Art Museum, San Antonio,** *Video Still, 2006*

1. WALKING

For the last few years artist Ernesto Pujol has dedicated his artistic practice to performative acts of intermittently walking and standing still. Because he imagines a world in need of repair, torn to shreds by loss and war, he attempts to catalyze its regeneration in small ways through the simple act of concentrated, focused, deliberate walking that engages mind, body and spirit.

The first of the series of Pujol's 'walks' began as a solitary act in the Magnolia Cemetery in Charleston, South Carolina, in July 2005. That original, unpremeditated gesture was an attempt to comfort the dead buried in this Civil War cemetery and to mourn the loss of life through war with the simple movements of a reflective contemplative *walk*. Pujol, who was a Trappist monk for some years before leaving that cloistered life to become an artist, walked among the long-ago deceased dressed in a monk's habit - a garment from his former life.

This gesture began a respectful series of slow mournful walking movements and kneeling gestures with only the camera as witness. The next walk, a more deliberate and public act of mourning, found Pujol as a 'character' of unknown origins at the McNay Museum in San Antonio, in June 2006, 'barefoot, pale in butoh white face, dressed in a long robe suggesting a mortuary shroud or a middle eastern garment', he wrote [unpublished], pacing the small circle before the museum's Kohler Fountain. This was a private mourning in the manner of the traditional monastic discipline of walking meditation or labyrinthine pilgrimage walks. But this particular walk was positioned as an art event that was visible to an art audience, who observed him from a balcony. At first the visitors paid little attention, but they were soon mesmerized by the focused seriousness of Pujol's intent. He has written that this walk was inspired by watching the daily count of US soldiers fallen in the Iraq war, which reached 2,000 in October 2005. He felt compelled to respond.

In his most recent and most sustained walk, in June 2007, he circumnavigated Spectacle Island outside of Boston. In this iteration, the walker has become the *Aguador*, the waterman, the water carrier. Here, as in many traditional pilgrimages, he has walked the island, taking samples of water back to the mainland. Dressed in a nineteenth-century garment, fragile as the action itself, he spent a day moving around the island followed by spectators originally assembled for an annual marathon run. While he collected vials of water samples, he was followed by a cohort of faithful witnesses. Water, Pujol has said, and not oil - or, rather, the unequal presence and absence of water - will probably be the next great crisis humans will have to negotiate. Water is essential, primal to human and animal needs, and it is dangerous to take its healing, life-giving, cleansing and quenching properties for granted. The 'waterman' collects small vials, specimens that serve as documentation for this simple action of sampling and preservation.

Here he saw himself in the role of the Bodhisattva performing the 'useless' task of *selling water by a river*. Why sell water by a river?

Perhaps so that those who often go near or on the river will come to *see* and *care for* the river once again. The *Aguador* creates a strange intervention, a vulnerable figure imagined from another time, alone amidst a crowd. 'The waterman will remain composed and silent at all times, as if in meditation,' Pujol writes, 'transforming a public moment into an intimate contemplative moment. His eyes will be closed when he is not moving. (He may only selectively respond to children.)' By the end of this performance, Pujol's antique garment had been battered by the wind and the sun. It was in tatters. And he had become part of the environment, attempting to melt into its space and time. His energy spent, he himself now was also in need of repair.

In these personas he has also walked the perimeters of fortresses, felt the interiority of buildings by running his hand deliberately and slowly across their skins. He had absorbed the animate and the inanimate, feeling the emotions of that which cannot articulate its pain. He has experienced that which appears fixed and inert, only to show us that it too is alive with memory and consciousness, if one can experience it as such.

These are ephemeral acts and interventions into the physical world with the intention of learning not about the otherness of the inanimate but about its sameness - that all forms of matter are one and that the living human body therefore can respond to them all, must respond to them - an old Buddhist belief that all matter has the potential to liberate and be liberated.

In his most recent iteration of walking, Pujol staged a sixteen-person durational piece in the rotunda of the Chicago Cultural Center in October 2007, where for 12 hours he and a group of young performers contemplated loss and manifested loss by the simple act of focused walking so that, by being the audience and the witness to their silent grief, we too could be healed.

2. STANDING

Kim Sooja, constructs offerings that become artworks and artworks that become offerings. She has sat silently in crowds in various parts of the world, with her arms outstretched, palms up, as in a *mudra* - 'A Beggar Woman' - silently asking help from passersby. She has stood still and silent while groups of people encircled her, gazing. She also has performed isolate actions only recorded on film. Stretched across an enormous boulder for hours on her side in Kitakyushu, Japan, in 1999, 'A Needle Woman' became an extension of the boulder while she silently contemplated nature. As the rock becomes part of her, she becomes the rock. She has not come to conquer the stone, but to enter into its vibration. Like Ernesto Pujol, she has merged with what would appear inanimate.

In 'A Laundry Woman' (2003), Kim Sooja stands

Kimsooja, A Needle Woman – Kitakyushu, 1999, *Single channel video projection, 6:33 loop, silent, Courtesy of the Artist*

with her back to the audience, completely still, contemplating water while the Yamuna River floats by. This place in India is downstream from a crematorium where offerings are made daily to the dead - flowers, paper lanterns, candles launched in memoriam. As we watch, the offerings are pulled along by the current of the river. 'A Laundry Woman' observes and witnesses the flow of the river, the force of life and the cleansing rituals of death. In India millions come to the Ganges each year to die in an auspicious and sacred place. Here, motionlessness is juxtaposed to the slow hypnotic movement of life's detritus. This spectacle conjures the notion of infinitude but also of finality. These offerings move off the screen to a place we cannot see. The river takes away all trace of life and death - the smallness and particularity of our time on this earth. There is no end to its motion, but there is an end to ours, and even an end to the blessings and the offerings of grief that well-wishers lavish upon us after we are gone.

In the embedded metaphor, one has to traverse the river in order to get to the other side, to cross the cosmic ocean to reach nirvana. Charon waits with his boat to take us across the Styx. Coins placed in the mouths or on the eyes of the dead were payment for his labour. We watch our own lives and our own deaths go by, as the figure stands with her back to us, simply, modestly, 'A Laundry Woman', watching as the remains of life and intent are cleansed. She does not determine the action, but she is its witness, using the physical body's capacity for stillness to transmit the spiritual.

Kim Sooja puts herself into situations often in potentially dangerous places - Cairo, Delhi, Chad, Rio de Janeiro, Lagos, Tokyo, New York, Mexico City, Shanghai, Havana. She is a witness to the intensity of place, yet not always a participant. When she stands silent in a dense crowd of moving people, her inaction is the enigma. She is 'A Beggar Woman', 'A Laundry Woman' or 'A Needle Woman' weaving her persona into the fabric of matter. On the edge of a giant rock, in Kitakyushu, Japan, 'A Needle Woman' creates unity with nature, maintaining stillness with the quietness of stone, expanding time. As Doris von Drathen has written, she is 'a needle, barometer, seismograph and compass ... indicating the everyday dramas that usually go undetected in our habit-formed lives' (2006: 37). She calibrates her body to the movement or lack of movement of the stone upon which she rests. In these still pieces, she sees herself providing an 'axis' for time and space, either vertical or horizontal, reminding us of the order of the world.

Ernesto Pujol traces the outlines of a wall. He feels the wall. And, if the wall is in trauma, he has been in trauma; he embraces the wall. Or he becomes the *Aguador*, the waterman. Both artists, stripped of pretension and of earthly identities, work to make the divine 'an aspect of the human world' - a prerequisite, according to Karen Armstrong, for the creation of archetype and myth (2005: 66).

3. SITTING LIKE A DUCK

How and when does the act of sitting become a gesture of provocation? Iraq-born artist Wafaa Bilal also puts his body on the line to remind us, as he says, that those who live in the 'comfort zone' do not understand 'the combat zone' (Bilal 2007). Here the artist *sitting* creates discomfort, as he makes himself a target for a world anxious to decathect its violence.

Under Saddam Hussein, Wafaa Bilal was labelled a political dissident. Scheduled for arrest and execution, he escaped into Kuwait and was then transferred to a refugee camp in Saudi Arabia, where he spent two years. During that time, he felt unsafe in the tents that were provided, and so as an artist he set to work each day making adobe bricks, drying them in the sun until gradually he had accumulated enough to build a small hut where he could both sleep more securely and teach art to the children of the camp during the day.

In the spring of 2007, working closely with other high-tech artists including Ben Chang, Wafaa Bilal chose to create a virtual war zone with real-life consequences in the Flatfiles

Gallery in Chicago. He titled the piece 'Domestic Tension', but had originally thought to call it 'Shoot an Iraqi', until he conceded that such a title might be too provocative.

Bilal's plan was to live in a room in the gallery for a month where he would eat and sleep and where, behind a Plexiglas shield, he could sit at his computer and maintain a continuous blog with those shooting at him. Participants watching him online via a live webcam positioned in the gallery could take aim and fire small pellets of yellow paint that would explode on contact. They could also engage him in dialogue, or they could do both - shoot at him and then verbally insult him, which many people did. It never occurred to Bilal that instead of the anticipated few shots a day, he would actually receive 40,000 shots in the first twenty days and a total of more than 62,000 shots, with people from 128 countries attacking him.

The paint gun was mounted on an armature that scanned the room in a robotic motion. When it found its victim it would open up and fire a shot so loud that it resonated like a real rifle blast, then it splattered paint on Bilal and all over the room. In the early days, before the device was perfected, paint balls flew out of the robotic gun with such velocity that they shattered the plastic shield. The shield had to be adjusted so that Bilal's physical welfare would no longer threatened.

The attackers were mostly video-gamers and paintball junkies 'intrigued by the possibility of shooting someone hundreds of miles away with a click of their mouse,' Bilal writes (2007). There were also bloggers who would hurl racist epithets and recriminations at him online if he went out of their sight for more than a few minutes. And there were those who kept him up all night shooting. But there were others. A group called Virtual Human Shield succeeded in jamming the site for seven days, keeping away those hackers who were trying to shoot at him continuously. There was Matt Schmid, a former US Marine who heard about the installation on the radio, went online, saw someone shoot at and break Bilal's only lamp, and came to the gallery the next day with the gift of a pole lamp, taller than the range of the robotic arm. There were those who brought food. There was a high-tech professional who heard about the piece and, anticipating the amount of virtual participation Bilal's web site would receive, came into the gallery and volunteered to connect the project to a larger server that could manage the unexpected volume of hits. He maintained the site for the entire run of the project. There were many friends and artist collaborators who helped Bilal develop the technology and actually build the device. They stayed close throughout. There was a very forward-looking, courageous gallery director willing to offer her white cube space for such an intervention and live with its destruction during Bilal's installation, when the room became a soup of splattered, sticky, smelly yellow paint. And there was Bilal himself, passionate and forgiving, whose attitude has always been that people simply need to wake up, to realize that the Iraqi

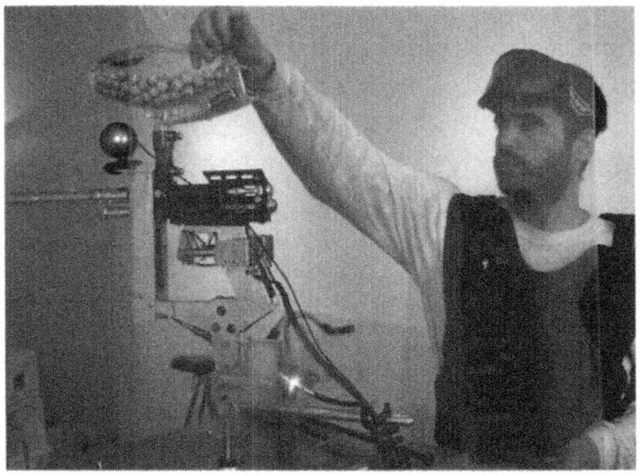

• **Wafaa Bilal, Domestic Tension 2007.** *Photographs by the artist.*

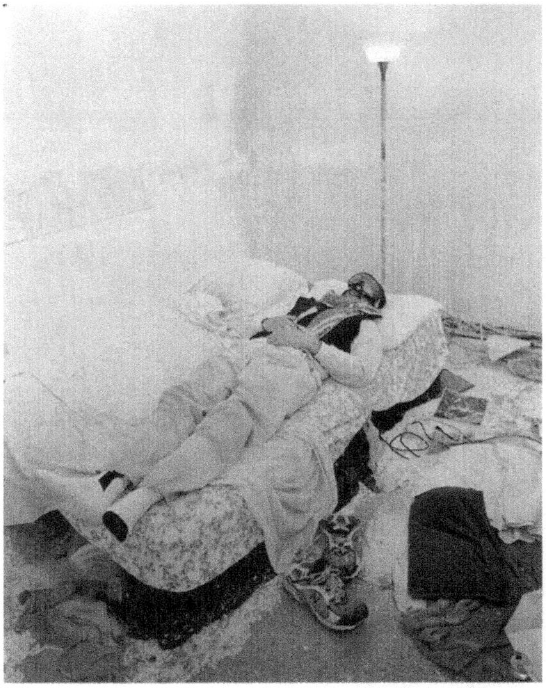

war is not a virtual war, not a video game, and that real human beings, with real names and real lives on all sides of the conflict, are being killed daily in Iraq. Bilal has said, 'Art doesn't have to change life, it just has to start something' (2007). No matter what people thought while entering into this encounter, they surely came out of it 'changed'.

Bilal positioned himself in the literal line of fire and waited. He did nothing while the world fought over him. In this he became representative of many things during his time online, but for most people his identity as *artist* was lost even though he positioned himself in a gallery and saw the entire action as performance - a deliberate inactivity of sitting still while the world took literal shots at him. Although it was a collaborative venture, he alone was the sitting duck. In the end he was so distraught by the gunfire, the lack of sleep, the randomness of the shots, the timing, the sounds, the no-escape, that he experienced the effects of post-traumatic stress syndrome, as if he had been in an actual war zone. It was astounding too that such conditions of war *could* be replicated in a gallery while the outer spaces housed regular art shows and on weekends were rented out to weddings.

Each of these three artists has created an axis of action to intercept daily life. Yet their interventions are modest given the enormity of their concerns - war, reparation, life, death, the passing of time, the development of human consciousness and responsibility. They simply point in the direction of their obsessions, sadness and impotence. Without actually meaning to, they come to reflect the unique ability of artists to engage the largest questions of life and society in their bodies, and to do so within mundane gestures of walking, standing and sitting - in full consciousness, yet without judgment. In their metaphoric embodiments and personifications of grave social concerns, they are unwilling to blame. In spite of their stated intentions, their actions actually render the rhetoric around most political concerns and activisms hollow and cowardly, because as humble as such performative acts may appear, they are courageous. These artists are willing to place their 'bodies on the line'. Nothing could be more dangerous or transformative, literal or metaphoric than this.

• **Wafaa Bilal, Domestic Tension 2007.** *Photograph by Christine Taylor.*

REFERENCES

Armstrong, Karen (2005) A Short History of Myth, New York: Canongate.

Bilal, Wafaa (2007) *Shoot an Iraqi*, ms. to be published by City Lights Press.

Caner, Daniel (2002) *Wandering, Begging Monks: Spiritual Authority and the Promotion of Monasticism in Late Antiquity*, Berkeley: University of California Press.

Drathen, Doris von (2006) 'Concrete Metaphysics: On the Work of Kimsooja', in *KIMSOOJA* (Stockholm: Magasin 3), p. 37.

Frye, Northrop (1969) *Fearful Symmetry: A Study of William Blake*, Princeton: Princeton University Press.

Russell, Norman, trans., (1980) *The Lives of Desert Fathers: The Historia Monachorum In Aegypto*, Oxford: Cistercian Publications.

Young-Bruehl, Elisabeth (2006) *Why Arendt Matters*, New Haven: Yale University Press.

The Liturgical Lens

MEGAN MACDONALD

Marina Abramović is a European artist whose performances are often associated with the spiritual.[1] She currently lives in the West, and her work is examined by Western academics. Abramović refuses to associate herself with any one spiritual practice, tending instead to engage with ritual and spiritual practices through performance alone, yet the comparisons made between her work and various spiritual practices are almost always non-Western.[2] In 2002 Abramović performed *The House With the Ocean View*, over twelve days in New York City.[3] Abramović's continuous performance took place in three suspended rooms, 6 ft deep, constructed in one section of the gallery. Each white box was hung next to but not touching the other. To the viewer's left was a room with a shower and toilet. The middle room held a table and chair and the room to the right had a bed, shelves for her clothes and a sink. Two other separate rooms were part of the event. One showed a video of Abramović and the ocean, the other was called the 'dream room' and participants could sign up for an hour slot to sleep/dream during the opening hours of the gallery. There was a line drawn across the room dividing Abramović's space from public space. People watched from any part of the room, either standing, sitting or lying on the floor. Abramović spent the time making eye contact with those who came as well as repeating actions such as singing, dressing, showering, drinking and rearranging the furniture.

The performance received critical attention from artists and academics, including Peggy Phelan (2004a, 2004b), Laurie Anderson (2003), RoseLee Goldberg (2004) and Thomas McEvilley (2003), who respectively compared *The House With the Ocean View* with Tibetan Buddhism, New Age, Australian aboriginal and shamanic/ yogic practices, Sufism, monasticism, contemplation and general spirituality.[4] None of these is necessarily inappropriate, but the absence of Christianity is conspicuous.

Despite what I see as obvious parallels with the Christian liturgy, references to Christian practices have been all but absent from responses to her work. This is not simply an oversight on the part of a few writers. Christian practices are rarely researched or written about in performance studies as knowledge of Christian liturgy has diminished, and with it the ability to identify and place manifestations of Western performances of ritual within a Christian paradigm has been almost entirely lost.

I would argue that this lack of attention and understanding has been informed by the powerful influence of anthropology, with its emphasis on participant observation in 'other' cultural systems, on Performance Studies. It was not until the end of the twentieth century that Western anthropologists began to examine the histories of their own cultures (Cannell 2006).[5] From its origins, anthropology looked outwards and Performance Studies (founded in the 1970s) followed this paradigm. During the foundational years of Performance Studies performance scholars used participant observation in

[1] The word 'spiritual' is used frequently and yet it is difficult to define one agreed upon set of meanings and uses. I employ it as it is used in performance writing, which is normally in relation to themes and uses of the body in performance. Actions or themes that recall ritual and/or religion, whether from the audience or artist's point of view are referred to as spiritual.

[2] Abramović was brought up in the former Yugoslavia, which has a long tradition of Christian Orthodoxy. While she herself does not follow one spiritual tradition, she has worked and spent time with various other cultures, learning about their philosophies and practices. She absolutely refuses to be associated with any religion or spiritual practice.

[3] This piece was offered to the people of New York as a response to the events of 11 September 2001. Abramović had conceived of a performance involving fasting and silence a few years before, but, as Sean Kelly says in his introduction to her book, the events of '9/11 changed everything profoundly for us all' (Abramović 2004: 5).

Performance Research 13(3), pp.146-153 © Taylor & Francis Ltd 2008
DOI: 10.1080/13528160902819489

studying ritual/theatrical/religious practices in cultures where the boundaries between various art forms and religion were more blurred than in the West (Schechner 1985, 1993; Turner 1982). By contrast, the study of religious rituals in Europe or North America has been the province of religious institutions and theologians. Where Christian practices have been examined in theatre or performance research, they are mostly discussed in relation to their *theatricality*, and not in relation to their belief systems, bodily or spiritual practices (Cremona 1998, Harrison et al. 2002, Harris 2003).

The participant-observer model provided information in two intimately connected ways: via an outside critical eye that maintained an almost scientific distance from the object of study, as well as from an inside subjective position that contributed, for example, feelings and embodied responses. For all its benefits in relation to the study of other cultures, this approach cast the culture in question as exotic or 'other'. While both anthropology and Performance Studies acknowledge the negative effects of othering in studying spiritual ritual practices (Bharucha 1993, Cannell 2006), few studies acknowledge that this methodology has also *privileged* non-Western rituals.[6] An outcome of this tradition is that Western scholars tend to explain the European and North American practices they do study in relation to other cultures, for instance comparing Abramović to everything but Western traditions. I do not propose that scholars should use comparisons with Christianity for the sake of merely balancing or correcting this practice. This would not solve the problem. The issue at hand goes deeper and extends to the influence of language in shaping how we understand the spiritual.

The participant-observer method developed in Performance Studies adopted a linguistic style closely akin to anthropology - with its slightly detached approach to description and evocative wording for personal experience. As anthropology strove to prove itself as a social science in the twentieth century through its discursive style, borrowing from it proved useful for the analysis of rituals/theatre from a performance perspective. However, the spiritually significant experiences from the cultures studied by performance scholars were sometimes transposed into the language of the scholar to detrimental effect. Scholars overlooked the way language itself functions in relation to concepts of belief, religion and spirituality. The interplay between a belief system and the language spoken by the believers is significant and must be taken into account by anyone seeking to understand the rituals associated with the belief. It is not just which words we choose but the meanings they convey that shape a specifically Western belief system. Spiritual language centres around beliefs, and it is this that poses the most significant challenges for research. The linguistic implications of belief produce a multi-layered experience which is, I want to argue, taken for granted by European-language speakers.

In 1982 the anthropologist Jean Pouillon identified a chronic problem that exhibited itself throughout anthropological case studies concerned with belief. 'Belief' was generally treated as a constant and compared with other cultures as such. Using another language as a comparison, in this case Dangaleat, which is spoken in northern Chad, Pouillon showed that while every meaning of 'to believe', can be translated into Dangaleat, the verb itself cannot. That is, the verb as it is used in French cannot be translated because there is no one equivalent term in Dangaleat. The unity that exists in western European languages, (for this pattern is also true of English, German etc.) is not evidenced in languages used by cultures whose belief systems are structured differently. Pouillon argued that Christianity is responsible for the idea of unity of meaning in the concept of belief. He focused on how structurally different Dangaleat and French are, as well as highlighting the influence of Christianity on shaping how the language of belief functions in the West.

The fact that the single term 'belief' needs to

4 This piece resulted in many articles, a book by the same name documenting the entire twelve days (and the use of her set in the television programme *Sex and the City*).

5 One recent publication has sought to redress the lack of research into Christian practices. Fenella Cannell's *The Anthropology of Christianity* (2006), contains twelve essays by anthropologists researching manifestations of Christianity from around the world. This work is welcome both for the scope of the essays and for the provocative introduction by Cannell, which challenges the role played by Christianity in the history, development and influences of anthropology. Performance studies would do well to follow this example and look again at Christianity as an important basis of Western performances of belief, as well as for how it has shaped current approaches to the analysis of ritual.

6 The participant-observer method does not, of course, preclude its use by those already part of a culture - a Corsican can study the rituals of Corsica - but its history has been predicated on the idea that there is more value in the 'other'.

be translated into a group of expressions in other languages raises the following question: what is it about belief which allows the meanings dispersed throughout many words in Dangaleat to be contained by one word in the West? Pouillon began by listing the most obvious of the usages:

> *croire à* - to state that something exists; 'to believe in the Devil' [this does not require the person to put their 'belief' in the Devil, only to believe in the existence of the Devil']
> *croire en* - to have confidence or faith; 'to believe in (trust in) God'
> *croire que* - for something to be represented a certain way; 'to believe that God exists'
> *croire en* - to give someone credit; 'to believe what someone says'
> *croyance en* - trust in a god such that there is a *credo*, 'a group of statements which become the direct object of belief'.
> (Pouillon 1982: 2-3)

The combined meanings show how complex a statement about religious belief in European culture might be, as well as highlighting the problem inherent in translating any discussion of belief into another cultural and linguistic context. Belief, in the examples above, can refer to existence, trust, confidence, credit or to a representation or the development of a creed. All these meanings can be understood by the use of one word as Pouillon demonstrates:

> *Croyance en* ['belief in, trust in'] God does imply *croyance à* ['belief in'] his existence, but implication is not identity. On the other hand, this implication seems so obvious that it often goes unformulated: a believer believes in [*croire en*, 'trusts in'] God, he feels no need to say that he believes in [*croire à*] God's reality; he believes in [*croire à*] it, one would say implicitly ... If I have confidence in a friend, if I belief in [*croire en*, 'have faith in'] him, will I say that I believe in [*croire à*] his existence? Certainly not; that existence is, simply, undeniable. It is only if it were not unquestionable that I would have to believe [*croire à*] it, and believe in it explicitly ... [I]t will probably be said that this is playing on words.
> (Pouillon 1982: 2, original formatting)

One term encompasses spiritual existence, the faith needed to relate to the spiritual, and communicates the representation of that relationship. Each individual who uses the term employs it in relation to some or all of these aspects, but it is impossible to 'know' which are meant in any one situation without questioning the utterer.

We in the West perceive a divide between 'us' and the spiritual realm. By comparison, the Hadjeraï of northern Chad - who speak Dangaleat - 'believe' in spirits called the margaï. Yet when the translations by early anthropologists from Dangaleat into English are examined, it is clear that their 'belief' is quite different than the Western distinction between perception, knowledge and belief:

> they believe in [*croire à*] the existence of the *margaï* like they believe in their own existence, in that of animals, things, atmospheric phenomena ... Or rather, they do not believe in [*croire à*] it: this existence is simply a fact of experience: there is no more need to believe in [*croire à*] the *margaï* than to believe if you throw a stone it will fall. One fears and/or trusts in them, one gets to know them, one gets used to them. (Pouillon 1982: 7)

While Dangaleat has words for 'believing', there is no unifying term that can be used to encompass all understandings and usages (1982: 5). The Hadjeraï have no need for one word, partly because their religion is based on what they see, feel and experience in the world around them; if there are spirits, then the spirits are just as much a part of the world as the people.

Pouillon's Chadean example highlights how European languages shaped what anthropologists reported about this group of people. The originating culture includes more possible actions in one expression (in this case the word 'belief') than does the destination culture (Hadjeraï). The kind of deity understood by the Western term God is personal and mono-theistic. Assumptions made by the originating culture take the idea reported by anthropologists that 'the Hadjeraï believe in spirits' to be related to the Christian idea of believing in God. However, these two belief systems cannot be compared point for point.

That there are different linguistic ways of expressing all that 'belief' entails is not surprising, but what the differences tell us is important in understanding our own religious traditions. Pouillon contextualized religious belief in the West thus: 'everything rests on a faith, which is simultaneously a trust and a specific *credo*. All the meanings of the verb 'to believe' should then come together, but this necessity is nothing more or less than a cultural necessity' (Pouillon 1982: 8). The traditions of Christianity, inform how we conceptualize belief and all that is spiritual, as well as how we use language to describe these experiences.

I want now to return to Marina Abramović and suggest that her performance *The House with the Ocean View* is closely connected with another neglected aspect of Christian practice - little analysed outside of theology - the liturgy.[7]
In fact, I want to suggest that her installation *is* a liturgy and to explore the reasons why.
The Christian liturgy brings together many of the issues discussed thus far, as it exists as a text, is performed by the participants and is both conceptualized of and performed as a spiritual event. The Roman Catholic liturgy is a set form of actions that centre around the Eucharist or Communion Rite - the blessing of bread and wine.[8] In the liturgy much is demanded of the body as it is required to sit, stand, sing, kneel, eat, pray out loud, pray silently and listen to music, text and teaching. The body smells incense, feels the sprinkling of water and reverberations of music, eats bread (on occasion also drinks wine), and shakes the hand(s) of friends and strangers. Actions are repeated many times within one performance of the liturgy and thousands of times by those who participate over the course of a lifetime. The priest makes possible an experience for the participants, sometimes acting as a stand-in, e.g., when the priest kneels at the high alter and prays, the participants kneel in their pews, as it would be impossible to build an altar large enough for everyone to kneel and pray together. At other times he performs actions which are only for him to accomplish, e.g., blessing the bread and wine and blessing the participants. The priest is not there to be solely watched as in theatre but to facilitate all other participants. He is not a shaman through whom participants gain access to the divine, yet he has trained for the position. The liturgy only really exists performatively - it has to be enacted in the space by the people who have come together with the express purpose of performing the liturgy; the written text in books is not the liturgy. Experiences of the spiritual, transformation and transcendence are all possible for those who undertake the performance of the liturgy. At the most basic level of structure there is a parallel to be drawn from the liturgy to Abramović's work - both use repetition of simple bodily actions to facilitate transformation.

In *The House With the Ocean View* Abramović offered a situation wherein the participants were taken out of the everyday and offered the chance to experience their own transformation through engaging with her experience. It was crucial to this performance that she live in the gallery 24 hours a day, because in contrast to the liturgical context, where the majority of participants know what is required of them, visitors needed guidance to become participants. The gallery had been set apart from the everyday through Abramović's decision to abstain from all but the most basic activities. In the liturgy a group comes together to create and perform the liturgy, and Abramović too was looking for this level of participation. However, as she was starting a new ritual, to succeed she needed participants and therefore had to establish a performance space that would allow anyone to engage. There was no training or familiarity possible. The actions she chose to include highlighted the demands of the body, hers as well as those of all who visited. Again, the main form of participation centred on performative action. To engage with Abramović by looking into her eyes was to enter into a simple yet unpredictable action that could only be accomplished in the moment by those present.

7 There are many styles of Christian liturgy. Among the oldest forms are the Eastern Orthodox, Coptic, and Roman Catholic.

8 The official teaching of the Roman Catholic Church is that the blessing of the bread and wine by the priest transubstantiates the elements into the body and blood of Jesus.

To sit or stand in the space with her was to perform the self simultaneously with her performance of self. Abramović did not offer a performance to be copied but facilitated a performance that would allow others to have their own performative experiences. Each person in the room had to relate the use of Abramović's body to their use of their own body. In relating the performance context thus, I have identified aspects of Abramović's work that can be directly compared with liturgical rhythms and actions.

While Abramović does not belong to any religious group and has often denied following one spiritual practice in her own life, she did grow up surrounded by two strong ideological systems - Communism and Eastern Orthodoxy. Her maternal grandfather was a patriarch of the Serbian Orthodox church who was murdered by the state for his beliefs. Her mother was a life-long member of the Communist Party and very committed to the cause. Abramović commented in a talk at the Frieze Art Fair in London in 2006 that it was the 'religious context of Orthodox[y] and Communism' together that was a large influence in one of her best known pieces *Thomas Lips* (Abramović 2006). There are, of course, differences between Roman Catholicism and Eastern Orthodoxy, between eastern Europe and western Europe, but the overarching spiritual traditions and linguistic histories in all these areas can be traced back to Christianity.[9]

In one of the many responses that address the spiritual side of Abramović's work, Peggy Phelan recounted her experience of staring into Abramović's eyes during the performance for what she thought was 10 minutes and finding out afterward that the staring had lasted for just over an hour:

> *All of this and much more passed between us. You looked and I saw, saw it all again. There was more too - but much of it still resists words. More than that, there are some things that should perhaps remain unsaid, because they are dangerous secrets and because they are truly mysteries* ... There was no object. There was a kind of fused subjectivity, a condensation of the main themes of psychic, emotional, and perhaps spiritual, development. It passed through and touched on aggression, surprise, trust, fear of betrayal, fear of annihilation, acceptance, connection, beauty, exhaustion, transformation. The strength of it still surprises me, not only because I remember it so vividly, all these months later, but also because at the heart of performance was an embrace of simplicity.
>
> (Phelan 2004b: 25)

Phelan's reactions to sharing Abramović's gaze include physical responses such as sweating and shaking, emotional responses of remembering personally significant events (such as September 11, 2001), and psychological/ spiritual responses of wondering what was prompting or enabling these embodied reactions. She talked of 'drifting away from [her] own consciousness', and of how she 'was not sure who [she] was becoming standing there' (Phelan 2004b: 25). These responses use spiritually relevant words such as 'transformation', 'trust' and 'mysteries'. Phelan stated that she had had a significant experience - but did not compare it with any personal religious/spiritual practices from her own life. However, in her academic analysis she used many examples of religious/spiritual practices, taken from outside the Western paradigm.

> Whether one calls it environmental theatre or social sculpture, *House* extends something of the repetition and serialization at work in Warhol's *Shadows* into the realm of live art. While Warhol was operating within the economy of the object and setting up repeating copies of the same image, Abramović was theatricalizing the repetitive everyday acts of sleeping, showering, eliminating waste and sitting at a table. But these acts, each perhaps an homage to the quotidian, did not render the performance a literal treatment of these common acts. On the contrary, the symbolic and metaphorical associations were dense, ranging from Kafka's Hunger Artist to the prayerful acts of a Sufi mystic. The accumulation of associations and meanings people brought to bear on the art quite literally added to its energetic force.
>
> (Phelan 2004a: 573-4)

9 Other religions have long been part of European society. This article only picks up on the influence in language and performance from Christianity.

While simultaneously showing how open the work is to individual identification, these comments also continue the othering I have discussed in relation to anthropological studies. The work may well bring to mind Sufism, but what will such a comparison mean to those who read the article in the West? Lack of knowledge about what a Sufi is and of how and why that is mystic serve to make Abramović's work mysterious. This comparison complicates more than it clarifies. But what could be put in its place? If Phelan had said 'Christian mystic', it would not necessarily have been crystal clear for an average reader, because any mystic practice is hidden, unknown and secret. However, the descriptive term 'prayerful acts' together with 'Christian mystic' should offer the potential for a Western audience to engage with what a prayerful act might entail, even if mysticism is not fully understood. It is not the use of Sufi in and of itself that I find problematic - but the constant use of examples that serve to exoticize a performance that can easily and profitably be framed by Western spiritual practices.

What is obvious to one person is not to the next. Another example of the uncritical use of language in documentation of the performance is provided by Phelan:

> Thomas McEvilley argues that Abramović's work 'is dedicated to preserving the traditional shamanic/yogic combination of ordeal, inspiration, therapy and trance'. Moreover, he astutely claims 'that this approach to performance art is both the most radically advanced - in its complete rejection of modernism and Eurocentrism - and most primitive - in its continuance of the otherwise discredited association of art with religion'. (Phelan 2004a: 21)

Besides the problematic use of term 'shamanic/yogic', McEvilley's claim that the performance is a 'rejection of modernism and Eurocentrism' makes me uneasy. The piece was, after all, performed by a European woman, to a predominantly Western audience in New York City and written about almost exclusively by Western academics and artists. McEvilley said that Abramović had enough knowledge and expertise to 'preserve' traditions in which she had no cultural history, limited training and to which she claimed no membership or long-term observance. Both McEvilley and Phelan list forms of spiritual engagement that are neither part of their culture nor part of Abramović's. Phelan uses McEvilley's writing to reference her own reading of the work yet does not define any of the terms more specifically, nor ask whether he was accurate in his analysis. McEvilley claims that Abramović's piece was influenced by - and is an example of - 'universal' rituals that transcend cultural distinctions across centuries if not millennia, as well as simultaneously crediting her with creating a new phenomenon hitherto unimagined in the history of performance. Without actually calling the piece an ideal marriage of art and spirituality, it was accorded the status of something capable of 'saving' people. Can it really be as old as the first people who expressed spirituality and radical enough to escape the bonds of cultural determination?

McEvilley also compares the piece directly with a vipassana retreat:

> *The House with the Ocean View* could be described as a meditation retreat made public. Specifically, it seems to have been based on what in the Pali tradition of Theravadin Buddhism is called a *vipassana* retreat. These retreats (which are given here and there around the world) usually last 10 to 12 days (Abramović chose 12), with no talking, reading or writing, and very limited eating; one can fast, as Abramović chose to do, or eat one meal at about noon every day. (McEvilley 2003: 117, 153)

In his article McEvilley goes into detail about the Pali tradition and another Buddhist tradition of meditation, linking Abramović's work to both. Here McEvilley indicates that Abramović did nothing but borrow an already existent meditation practice and put it in a gallery. The major change that she incorporated was looking directly at people, which would not normally be part of a silent retreat. Yet the overarching idea is that Abramović accomplished something quite incredible and unique. This is the most thorough and convincing comparison of her work with any one spiritual practice that I have read. Given, however, that even McEvilley chose to use a range of practices as points of comparison in his article, I still find that something is missing. When I look at the same basic facts: a twelve-day meditation, fasting and silence, I think of Christian retreats, monasticism and ascetic practices. The number twelve is also relevant. It echoes the twelve days of Christmas, twelve disciples of Jesus and twelve tribes of Israel to name but a few. Abramović mentioned that twelve was important to her because the number three was a significant number and 1+2=3. Again three is a number of importance in Christianity, not least because it is used for the trinity of God the Father, God the Son and God the Holy Spirit. McEvilley was not wrong to explain something of vipassana, but where are the other comparisons to traditions closer to Western culture?

In another section Phelan linked spiritual practices with concepts of eternity:

> Abramović, who has been deeply influenced by Tibetan Buddhism and shamanic wisdom from disparate traditions, learned during the early 1970s that the border crossing traversed within performances that work on the art/life divide might be seen as a kind of rehearsal for that other crossing, the one between life and death.
> (Phelan 2004b: 19)

There is nothing surprising in the idea that artists are interested in the themes of life and death. However, Phelan revealed some implicit assumptions about how belief is constructed. The concept of crossing from life to death supposes a certain kind of construction of the soul: that we have souls, that when we die our souls go somewhere else, and that life and death are separate stages of experience. The two spiritual practices included in the quotation are explicitly linked with life and death but are not similar to each other in their understandings of life and death. There is not room here for a detailed discussion of how Tibetan Buddhism approaches reincarnation or of how the hundreds of types of shamanism deal with death and spirits, but neither set of practices assumes absolutely that a soul has one life and then passes on to eternity, as does Christianity. Neither assumes a mono-theistic God, or that the soul can have a personal relationship with a god, or that the soul is linked to an individual who will remain that individual for all eternity. The idea that a soul can cross from life to death implicitly carries within it the understanding, at least in the West, that the soul leaves the body and travels to heaven, hell, the universe in general or (in more recent history) just to nothingness. That Phelan used a phrase completely related to belief but did not more closely define how those differing belief systems operated in relation to life and death is an example of Pouillon's argument. He argued that Western language and religious history are intertwined such that the languages cannot but reproduce the logics of the Christian influence of the last two thousand years. To compare the practice of Western religion with any other forms, and to do so in a Western language, is therefore to risk blindly applying the logics of Christian spiritual practices.

Western scholars have almost lost the ability to read the culturally specific Western references in performance art that is spiritual. When confronted with a piece like Abramović's, which contains many convincing points of comparison, it is almost as if a blind spot exists blocking out that which is most familiar. Or perhaps the Christian traditions have become so unfamiliar in contemporary Western societies that researchers are simply unaware of what they are watching. Either scenario needs to be rectified; we need to be aware of the history of our own traditions and of how they continue to shape us today. This is but one example of how to apply what I am terming a 'liturgical lens' to performances that have spiritual content. What is needed is a renewed awareness of the history of Western practices so that knowledge about rituals and their spiritual significance is not lost.

REFERENCES

Abramović, Marina (2004) *The House with the Ocean View*, Milan: Charta.

Abramović, Marina (2006) 'Seven Easy Pieces or How to Perform', Frieze Talks (Frieze Art Fair), London: Frieze Foundation, <http://www.friezefoundation.org/talks/detail/seven_easy_pieces_or_how_to_perform/> (accessed 20 October 2006).

Anderson, Laurie (2003) 'Marina Abramović', in *Bomb Magazine 84* <http://www.bombsite.com/abramovic/abramovic.html> (accessed 17 November 2006).

Bharucha, Rustom (1993) 'Collison of Cultures: Some Western Interpretations and Uses of the Indian Theatre', in *Theatre and the World: Performance and the Politics of Culture*, New York: Routledge, pp. 13-41.

Cannell, Fenella, ed. (2006) *The Anthropology of Christianity*, London: Duke University Press.

Cremona, Vicki Ann (1998) 'Re-Enacting the Passion during the Holy Week Rituals in Malta', *Theatre Annual: A Journal of Performance Studies* 51: 32-53.

Goldberg, RoseLee (2004) 'The Theater of the Body', in Marina Abramović *The House with the Ocean View*, Milan: Charta, pp. 157-9.

Harris, Max (2003) 'Saint Sebastian and the Blue-Eyed Blacks: Corpus Christi in Cusco, Peru', *TDR: The Drama Review* 47 (Spring): 149-75.

Harrison, Paul Carter, Walker II, Victor Leo and Edwards, Gus, eds (2002) *Black Theatre: Ritual Performance in the African Diaspora*, Philadelphia: Temple.

McEvilley, Thomas (2003) 'Performing the Present Tense', *Art in America* (April): 114-17, 153.

Phelan, Peggy (2004a) 'Marina Abramović: Witnessing Shadows', *Theatre Journal* 56: 569-77.

Phelan, Peggy(2004b) 'On Seeing the Invisible: Marina Abramović's *The House With the Ocean View*', in Adrian Heathfield (ed.) in *Live*, London: Tate Modern, pp. 16-27.

Pouillon, Jean (1982) 'Remarks on the Verb "To Believe"', in Michael Izard and Pierre Smith (eds) *Between Belief and Transgression: Structuralist Essays in Religion, History and Myth*, trans. John Leavitt, London: University of Chicago Press, pp. 1-8.

Schechner, Richard (1993) *The Future of Ritual: Writings on Culture and Performance*, London: Routledge.

Schechner, Richard (1985) *Between Theatre and Anthropology*, Philadelphia: University of Pennsylvania Press.

Turner, Victor (1982) *From Ritual to Theatre: The Human Seriousness of Play*, New York: PAJ.

Wholly Unholy

Religious iconography in Israeli art and performance

SHARON ARONSON-LEHAVI & NISSIM GAL

A group of twelve actors dressed in black sits scattered among the audience which surrounds an empty, brightly-lit platform. An actress gets up from her place within the audience and slowly starts walking in circles on the platform, quietly singing/reciting a few lines from Psalms: 'My heart is sore pained within me; And the terrors of death are fallen upon me. Fear and trembling come upon me, And the horror hath overwhelmed me' (Psalm 55: 5-7). This is the opening scene of Rina Yerushalmi's eight-hour epic Bible Project (1995/1998), in which she selected and reedited about eighty excerpts from the Bible (Old Testament) but kept them in the original poetic Hebrew.[1] Throughout the performance the twelve actors recite the texts in the third person, enabling the most familiar stories - along with unexpected biblical texts such as lists of laws and genealogical lists of fathers and sons - to filter through the performers' bodies and come out anew, offering fresh perspectives on the ancient texts.

Whereas one would expect such an epic performance to open with the scene of the creation of the world and with God's presence and omnipotence (like medieval mystery plays, for example), the opening lines of the performance demonstrate the opposite. In this scene, as in the rest of the performance, the perspective is human, a woman who expresses her all-too-human fear of death and finality. As she enters the stage from the audience, she functions as a secular cantor, representing herself as an individual person and as a member of the audience from which she has just come. Walking in circles evokes simultaneously a contemporary existential image of wandering and a ritualistic pattern of continuation. Thus spectators find themselves intimately tied to the rhythmic language and context, experiencing a sense of belonging and at the same time a defamiliarized moment of detachment.

For an Israeli (Jewish, Hebrew-speaking) audience, intimately familiar with the Bible as text, foundational myth and language, Yerushalmi's performance evokes a 'wholly unholy' experience in its simultaneous use of ritualistic performative elements and deconstructive interpretations. Throughout the performance Yerushalmi juxtaposes the biblical texts with theatrical interpretations that bring out traditionally silenced voices, for example performing the scene of the flood from the perspective of the drowning people rather than that of Noah (Fig. 1), or Isaac's miraculous birth through the eyes of Sarah, an old woman lying on the floor suffering the pain and hardship of labour (Fig. 2). By focusing on the human and other point of view, the performance uses the status and power of the biblical texts to criticize contemporary society and claims responsibility for social and national issues on the people who live in the present and who are part of the audience.

This performance reflects in multiple ways the complex connection between Israeli society and religious culture. Rooted in the Zionist project, Israeli society exists on the threshold of the

[1] *Va Yomer. Va Yelech. Va Yishtachu. Va Yare.* (And He Said. And He Walked. And They Bowed. And He Saw.) Performed by the Itim Ensemble in affiliation with the Cameri Theatre of Tel Aviv. Created and directed by Rina Yerushalmi (1995/1998), <http://www.itimtheatre.com>.

Performance Research 13(3), pp.154-162 © Taylor & Francis Ltd 2008
DOI: 10.1080/13528160902819505

• Fig. 1. Rina Yerushalmi (director), Itim Ensemble, *Va Yomer. Va Yelech* (And He Said. And He Walked) 1995, The Flood, drowning victims. Left to right: Iyar Wolpe, Lior Ben Avraham, Carmit Mesilati, Eran Sarel, Emanuel Hanon, Moisi Shmuel, Noam Ben Azar. *Photo by Gadi Dagon.*

religious and the secular. Although Zionism had a vision of a new society - emphasizing secular aspects of Jewishness - it nonetheless derived its ideological power from its emphasis on a Jewish identity, the relation of Jewish people to the Land of Israel, the Bible and the revival of the Hebrew language as a spoken language (see Aronson-Lehavi and Rokem, 2009 [forthcoming]). In a performance that embodies this duality and tension between secularist and religious culture by refusing any traditional or religious iconography of biblical representations (no palm trees), there is only one scene in which the Bible as book appears on stage. In Scene 9, 'If thou shall hearken unto the voice of the Lord thy God to observe to do all his commandments,' (Deuteronomy 28), another actress stands alone on stage holding in her hand a small, black Bible (Fig. 3).[2] This emblem, the small black book, is used not only in religious but in secular contexts, and therefore evokes associations with modern Israeli existence: the small black book is studied in the secular education system, given to students and soldiers at high-school and martial ceremonies and is associated with the *raison d'être* of the Zionist vision of a new Israeli culture. Holding the closed book in one hand as the actress does, turns it into a trivial yet powerful, daily yet ritualistic icon. Each member of the audience is tied to it, each in a different way. From the perspective of an Israeli theatre-goer, this performance simultaneously reinforces a sense of a congregation, yet - as in the rest of the works of art we will go on to discuss - deconstructs the idea of a congregation by individualizing each person's relation to religion and critically examining contemporary social questions through religious icons.

In the following pages we examine how central issues within contemporary Israeli discourses - nationalism, militarism, feminism, poverty and human/minority rights - are critically negotiated in art, theatre and performance through adopting religious iconography and exploring the tension between the sacred and the profane. We propose the possibility of a postmodern congregation, examining works that exist on the threshold between holy and unholy signifying systems.

• Fig. 2. Rina Yerushalmi, Itim Ensemble, *Va Yomer. Va Yelech* (And He Said. And He Walked) 1995. Sarah giving birth to Isaac. Titina Asepa and Moisi Shmuel. *Photo by Gadi Dagon.*

[2] A Bible book in Israeli context refers to the Old Testament and does not include the New Testament.

• **Fig. 3. Rina Yerushalmi, Itim Ensemble, *Va Yomer. Va Yelech* (And He Said. And He Walked), 'If thou harken unto the voice of the Lord thy God' (Deuteronomy 28), Neta Yashchin.** *Photo by Gadi Dagon.*

POSTMODERN CONGREGATIONS: TOWARDS A THEORETICAL DEFINITION

In Hebrew the words *kahal* (audience) and *kehila* (community/congregation) come from the same root: ק-ה-ל (k-h-l), which means 'gathering'. Differentiating between these two social kinds of gathering - being a member of an audience or of a congregation - is made possible through examining the different functions of these two concepts. Whereas gathering for the purpose of a ritual (religious or other) is meant to lay emphasis on a group's collaborative intentions and wishes/needs to confirm and reinscribe an identity and a sense of communality, the audience of a theatre performance or art exhibition is usually understood as a group of individuals which might or might not share social common denominators. Accordingly, a performance or exhibition potentially aims to subvert and deconstruct norms and values, attempting to free the audience member from the characteristics and definitions of the role of a congregant.

A unique site in contemporary art and performance that allows for the blurring of distinction is performance that directly addresses and performs religious icons and texts, using their embededness and 'congregational' appeal on the one hand while performing them in deconstructive and 'unholy' ways on the other. In the Israeli performances and works of art we examine in this essay, a third, hybrid kind of 'gathering' is evoked, which we term 'wholly unholy'. As opposed either to a community/congregation in the traditional sense - one that gathers around a religious sign system in order to define its identity as whole and adhere to a certain definition - or to an individuated audience that gathers to watch performances in which religious symbolic systems are denied meaning, we suggest a third kind of a congregation, one that combines communality and individuality, one that gives up wholeness but not holiness/meaning.

Jerzy Grotowski, like Antonin Artuad before him, identified this need of the modern, secularized society:

> The spectators are more and more individuated in their relation to the myth as corporate truth or group model, and belief is often a matter of intellectual conviction. This means that it is much more difficult to elicit the sort of shock needed to get at those psychic layers behind the life-mask. Group identification with myth - the equation of personal, individual truth with universal truth - is virtually impossible today.
>
> (Grotowski 2002 [1965]: 23)

However, although creators such as Grotowski sought a new kind of a community, established through acts of profanation and 'confrontation' as in his 'poor theatre', allegedly turning an audience yet again into a *congregation*, the idea behind such a theatre still implies the possibility of defining a community, one that is whole, one that adheres to 'universal truths'. Grotowski seeks 'an experience of common human truth' (23).

Such utopian ideas regarding the redemptive potential of theatre and art on the one hand and the possibility of turning an audience into a universal congregation on the other tend to neglect difference and ignore the existence of multiple communities, ideologies and identities that might exist within any social framework.

By looking at contemporary Israeli art and performance art that might be considered secular and potentially subversive, we suggest that individuating the collective symbols opens up the option of representing many perspectives and of creating a heterogeneous map of congregations. By turning to and performing elements taken from religious and mythical discourses that originally define unified identities, which are based on similarity, such performances enable us to imagine the shared and the common from the perspective of difference and otherness without giving up belongingness. It is significant to note that Israeli artists rework Jewish and Christian religious iconography. This is especially the case in relation to performances and works of art that centre on the theme of the sacrificial figure.[3] In order to demonstrate this idea we have chosen examples from several artistic mediums: photography (Adi Nes), video performance (Sigalit Landau), theatre (Rina Yerushalmi) and painting (Asad Azi). The following examples are part of a larger Israeli performative and artistic discourse that negotiates social and political issues though reinterpreting religious iconography, both Jewish and Christian.

RE-PLACING THE SACRIFICIAL FIGURE

The theme of the sacrificial figure is central to almost any religious thought system, and it has acquired a central place in Israeli culture. Two kinds of sacrificial figures characterize Israeli discourse - the post-Holocaust Jewish identity and the Israeli soldier, a hero who bravely gives his life in defending the country - most often allegorized as Isaac being sacrificed by Abraham. However, in the texts we now examine, created since the 1990s, a discernible 'opening up' of this category has occurred, creating space for other kinds of sacrificial figures to be heard, problematizing the very idea of the necessity of a sacrificial pattern and demanding a new kind of social responsibility.

Adi Nes's photography uses the absolutist dimension of religious art in order to critically examine the local ground of Israeli society and culture, confronting transhistorical religious truth with a concrete image of social injustice. His well-known 'Soldiers' series created between 1994 and 2000, which included his photograph of thirteen soldiers sitting and eating at a long table (Untitled [1999]) - clearly and ironically referring to the image of the Last Supper[4] - offers a new perspective of the sacrificial figure in the guise of the Israeli soldier. The soldier is considered a signifier of the national Israeli community that perceives itself as a society in permanent risk of war and struggle. In *Untitled* (1995) (Fig. 4), the rectangular format is crossed by the weak body of the wounded soldier, his right hand dropping on his side and his upper body lying on his fellow soldier.

While the soldier lying down displays a horizontal field, the other soldier posits a vertical

[3] There are many reasons for the presence of Christian iconography in Israeli art, among them the attempt to create works with a 'universal' appeal that relates to a Western, Judeo-Christian cultural heritage.

[4] These images are reproduced in Perez (2003).

presence as if lamenting or taking care of the wounded soldier on the ground; the soldier's wounds are explicitly marked on his side and hand, echoing the wounds of Christ and blood drips out of his mouth. This battleground scene is forcefully highlighted and contrasted dramatically with the darkness surrounding it. The military couple is situated in the moment, in the field of the 'real', by being exposed and lying on the bare, stony, thorny land; the circular thorny area is highly significant and forcibly indicates Jesus' crown of thorns.

Nes's picture is based on, among other sources, the medieval theme of the *pietá* - depicting Mary mourning over the dead body of her son lying on her lap.

In medieval thought this posture of the lamenting Mary was interpreted as a parallel theme to the crucifixion itself, the two themes considered a-historically as imaginary keys to understanding revelation through the divine and the earthly. Nes appropriates the religious formula and customizes it to his needs. While Mary is the lamenting figure in the conventional versions of this theme, Nes reverses the gender of the figures and posits a male performer instead of holy mother. This act is not innocent. It relates to artistic precedents that comment on the patriarchal tradition of Western culture, because the critical effect of this gender-crossing is the elimination of what might be formulated as the traditional division between the female figure as the origin of sin and the male figure as that of salvation.[5]

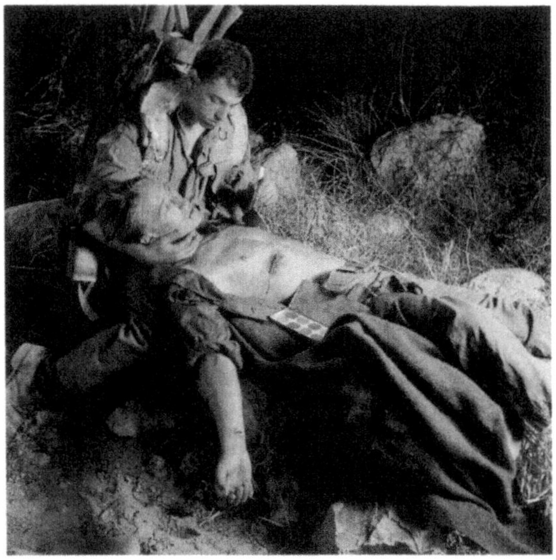

• **Fig. 4. Adi Nes, *Untitled* (1995).**

Nes's photograph exhibits the way passion and sexuality were traditionally channelled into the frame of the dominant ideology. In the Israeli context, sexual liberation and the apparent, manifest body were connected with the new society being built. However, Nes's photograph fluctuates between the structured poles of the visual discourse of the homosocial and the homosexual. The heterosexual patriarchal presupposition is that there can be a clear division between homosociability and homosexuality.[6] Nes's image charges the Christian theme with homosexual connotations that invite us to rethink the familiar images of the pietá and the battlefield in the discursive frame of desire, according to which homosocial relations conserve in themselves an erotic potential and do not enable its repression. The masculine scene prevents the possibility of viewing this image of the soldier as a strong undefeated or invincible figure. On the contrary, the position of the two men presents the option of masculine vulnerability, as the male body is exposed in suffering and torment. The relations implied between the injured figure lying on the ground and the figure treating him are displaced by the gaze of each character. While the military medic casts down his eyes, the wounded soldier raises his and examines the medic, hence the beholder cannot ignore or repress the complicated economy of desire between these two men.

Whereas the Zionist discourse tries to constitute a masculinity that represses feminine marks that might threaten a harmonious ideal of the muscular body, Nes parodies this ideology in his photographs by highlighting the intimacy and physical attentiveness between the men. The aestheticization and the gendered complexity Nes injects into the Christian theme is an explicit criticism directed at the militarism of Israeli society. In addition, the universalistic

5 For this formulation in the context of art, see Meyer (1997: 20-1).

6 'Homosociality' refers to same-sex relationships that are not necessarily romantic or sexual, i.e., a relationship between heterosexual males can be homosocial. Eve Sedgwick (1985) claims that the differentiation between the homosocial and the homosexual is instrumental and might express homophobic ideology, while actually there's a continuum (not an identity) between the two.

conceptualization of the body in terms of sublimity and redemption is displaced; the physical body is exposed through challenging the ideological subjugation of desire to the national project. The horizontal soldier's sloping hand, his wounded body and his weakness correspond with the figure of the living-dead. The posture and the gaze of the wounded soldier lead us to discover the jouissance that permeates the scene; the torment of the body and coming death are not an expression of a national revelation: this is not a self-sacrifice for the benefit of the state but an experience of non-collective pleasure. Contrarily to perceiving the dying soldier as the necessary sacrifice for the creation of a collective national image, we are offered a concretization of the experience of desire between men, a sexuality that is not subordinated to the law of the nation. In other words, while deconstructing the conventional iconographic depiction of the pietá, Nes uses this very formula to express ideas of tenderness and care that are part of this icon.

In addition to Nes's reappropriation of the themes of care and desire, he also explores through the image of Christ's sacrificial figure the theme of the wounded body. The incision on the right side of the soldier's body explicitly refers to Christ's wound in Caravaggio's *The Doubting of Thomas*. Caravaggio renders Thomas's drama of disbelief with shocking effect as the characters in the painting look for the truth the beholders cannot bear to confront. In Caravaggio's painting the distortion of the body is connected to the urge to constitute the real as a stupefying experience, to echo art historian Michael Fried (1985: 73). In this realist tradition the act of seeing, i.e., literally depicting the scratching or touching of the wound, is translated into the act of wounding the eye and effacing the seeing by the act of looking. In contrast to Adi Nes's image, we almost cannot look at Caravaggio's depiction, because the confrontation with the violating act of Thomas is too painful, the wounding of the eye in fact causes one to 'look away', catching the spectator in process of wounding. Thomas's touch is contradicted by the way his gaze 'misses the point', slips beyond what he touches. This 'refusal to look' is signified in Nes's 'makeup' wound, signifying once again a certain lack in culture, the void in the politics of visibility. Nes suggests that in the present culture there is no place for refusing to look and see and accordingly demonstrates how violence is sealed in the gesture of applying makeup to the wound. The hyper-visibility of the makeup as a pleasant image suggests on the one hand an expression of a political stand criticizing the repression of minorities, yet on the other it digests all the invisible dimensions of reality turning the effect of the real into the 'total makeover' of the cliché, emphasized by the makeup brush the medic holds in his hand.

In apparent opposition to Nes's use of artifice, Sigalit Landau's video-art piece Barbed Hula (2002) also forces us to look at the real wounded body of a sacrificial figure. In this video the artist herself stands naked on a seashore, injuring her body as she plays hula-hoop with barbed wire. In this performance piece she combines religious narrative with historical memory: the thorny hula-hoop evokes Christ's crown of thorns, while the barbed wire creates a clear association to the fences that surrounded the death camps during the Holocaust. By turning her own body into a feminine image of a Christian figure that also bears traces of the Jewish traumatic history, Landau demonstrates an idea of universal pain that exceeds time, place and religion. At the same time, in her performative act of bringing together Christian and Jewish icons of pain and victimization, Landau individuates this pattern since the barbed hula evokes not only an image of suffering but also of bodily pleasure and even nostalgia. Like a child playing with a hula hoop, Landau's naked body on the seashore - which gets progressively more injured - symbolizes on one hand a history that cannot be erased and a genuine, primeval search for a renewed naïveté on the other, while her actual pierced and injured body disables any possibility for such naïveté to be realized.

• Fig. 5. Adi Nes, ***Untitled (Hagar)*****(2005) in *Bible Stories* (2003–7), *Hagar*.**

In *Standing on a Watermelon in the Dead Sea*, another video-art piece by Landau, presented as part of her exhibition *The Endless Solution*,[7] she once again combines an Israeli identity with the Christian theme of the sacrificial figure, and once again offers a feminist interpretation of this image. In this video piece, Landau balances herself on a watermelon in the Dead Sea in the south of Israel. She stands naked with her arms outstretched in the shape of the crucifixion. Because of the density of salt in this sea, the watermelon, despite its weight, floats and therefore pushes Landau's body up, while she, by forcefully balancing herself on it pushes it back down into the sea. This dual act of pushing up and down, while performing a cross-gendered Christ figure, demonstrates the paradoxical move towards and from religious signifying systems. Landau's aspiration for freedom - she looks like a bird trying to fly and yet strongly holds onto the watermelon that symbolizes her femininity - is negated by her inability to let go of cultural heritage, history, geography and religious iconography. In this video performance, as in Barbed Hula, although Landau's identity is determined by gender and cultural and mythological history, as in Nes's image the relation to iconography is individuated and makes place for the performer's own body and desire. While Landau's body refers to the Christian image of the sacrificial figure, in the central hall of the exhibition a huge construct of a head of a ram lies on the floor, metonymically signifying the myth of binding Isaac. It is around this centre that other stories of victimization become present.

The contemporary political situation of the Israeli-Palestinian conflict is also examined by many theatre creators and artists through religious iconography. Asad Azi's *Woman and Flag*,[8] for example, critically weaves religious symbolism into nationalism. The painting has two layers: a tender woman wearing a colorful, light and floral garment holds her son tightly, staring at the beholder. On that tenderly painted image of the 'Madonna and Child', there is a black graffiti painting of the Israeli flag and the word 'Golani' (an IDF brigade) in Arabic letters. The graffiti of the national symbol on top of the religious one is an act of erasure. It refers to acts done by Golani soldiers who were fighting in Gaza during the first Intifada in the 1980s. Whereas the 'Madonna and Child' is an expression of daily life, the graffiti destroys the possibility of such an aesthetic existence. By opposing the national symbolism with the religious one, Azi uses the religious iconography to criticize the violent act. In this context the religious schema has a performative power.

RETELLING THE PRESENT: BIBLE STORIES

Like Rina Yerushalmi's Bible Project, Adi Nes's *Bible Stories* (2003-7) is a collection of photographs based on biblical characters and stories. The images and locales include a soup kitchen, women picking food in the garbage, homeless men, beggars, and men fighting violently. Nes creates a seemingly journalistic documentary project with social motivation, trying to bring to the front stage an awareness of the difficult reality the bourgeoisie seeks to

7 Sigalit Landau, The Endless Solution, Helena Rubinstein Pavilion for Contemporary Art, Tel Aviv Museum of Art, 23 December 2004 - 30 April 2005. Curator Mordechai Omer. 'The Endless Solution' refers to the Nazi term 'the final solution' and creates in Hebrew a play on words.

8 'Flag Nation Flag' Exhibition, Center for Contemporary Art, Haifa, 1998.

repress: a report on poverty. Nes stages the biblical figure of Hagar as a contemporary woman sitting on a neglected, urban staircase, dressed in rags, covering her mouth and stretching her hand as if she were a beggar (Fig. 5). This photograph suggests the iconic photograph by Dorothea Lang from 1936 taken during the great depression in the United States. Nes's woman/Hagar is a lonely figure without her son Ishmael, who gave her the status of a biblical hero and was the reason for her banishment from the house of Abraham.

In another picture the artist depicts Ruth and Naomi as contemporary gleaners echoing Millet's famous composition (1857). These two women collect the leftovers in the margins of the market.

The difficult visibility of poverty exposes Jewish ethical concepts that have fallen into oblivion in contemporary society: taking care of the stranger, covertly giving charity, donating a tenth of one's crop, hospitality, etc. In these contemporary biblical images Nes asks us to reconsider the impossible reality of people whose status as a minority or whose 'marginality' exposes the teeth of capitalism, ethnocentrism, patriarchy and the society of the spectacle. The juxtaposition between the people who wander on the margins of society lacking any expectations or future with the heroes of the Bible challenges the concept of the 'chosen people'.[9] In this collection - as in his previous artistic projects - Nes also appropriates compositions and themes from the religious Christian art. However, in his deconstructive interpretations he takes the position not of an orthodox interpreter but of a preacher.

Rina Yerushalmi also stages the biblical texts in ways that comment on the social and political present of Israeli society. The second part of the performance *Va Yelech* (And he Walked) depicts Israel's/Jewish recent history through the Bible. Yerushalmi theatrically suggests that the Exodus and the biblical conquering of the Land of Israel have been repeating in modern times. The parallelism between the post-Shoah wanderings in the 'desert' and the immigration to Israel then and now, is theatricalized as the twelve actors

[9] In addition to the figures discussed here of Hagar, Ruth and Naomi, Nes presents in this series the figures of Noah, Abraham and Isaac, Job, David and Jonathan, Samuel and Saul, Elijah, Joseph, Jacob and Esau, and Cain and Abel.

• Fig. 6. Rina Yerushalmi, Itim Ensemble, *Va Yomer. Va Yelech* (And He Said. And He Walked) 1995. Left to right: Neta Yashchin, Noam Ben Azar, Lior Ben Avraham, Zahala Kuras. *Photo by Gadi Dagon.*

represent the twelve tribes, reciting the long list of the rest-camps in the desert (Numbers 33) in twelve different languages simultaneously: each performer using the language of the country from which his or her own parents came to Israel. In addition, all twelve actors stand in a row and hold brown suitcases that unquestionably signify the Holocaust (Fig. 6). This scene, which leads to the building of the country as the group of Holocaust survivors gradually becomes a strong martial force, uses the biblical lanuage in an unexpected way. In a performance that celebreates the poetic biblical Hebrew, the fact that the twelve actors recite the list of stations in twelve languages at once creates a babylonian gibberish and sense of disorientation. Rather than an image of a unified Israeli identity, this scene opens up this concept and demonstrates the existence of multiple identities. In addition, rather than signifying the 'people of Israel', the 'congregation', this scene performatively individuates the performers' own identities and relation to the Bible, heritage and Israel.

Following the scenes that depict the Holocaust and the wandering in the desert, the group fronts the next stage that is a necessary part of building the state - the wars. Here Yerushalmi demonstrates the painful price of the wars as the performance platform slowly becomes a graveyard, geometrically covered with cards on which biblical verbs are printed. While the striking scenes of war and sacrifice lay responsibility, express and problematize a very concrete reality of the present, Yerushalmi paradoxically asserts that the biblical mythological narrative is repeating itself. In the final analysis the historical and mythical power and narrative rise and project themselves above the surface.

CONCLUSION

The idea of postmodern congregation that we have tried to articulate here is closely related to contemporary discourses of individuality, the body and identity politics in Israeli society. Although these artists do not represent a self-declared movement or create manifestos as in modernism, they all work within a shared conceptual framework. Rather than suggesting any sort of contemporary collectivism, we are trying to point out the existence of many different communities and ideologies that nonetheless share a deep cultural common denominator.

All the works of art discussed simultaneously tie themselves to canonic and iconic texts and images and try to express multiple identities through them. The common denominator has to do with how they all reuse and reinvent the texts and contexts of the mythical and religious texts and icons that they use. The tension between the secular and the religious becomes a place for exploring contemporary social, moral and ethical issues. A work of art that is 'wholly unholy', or rather holy un/wholly, uses the tension between the religious and the secular, or the sacred and the profane, as a productive source of inspiration, finding in this space of tension a renewed creative energy that does not nullify religious truths but enables the voicing of traditionally silenced ones.

REFERENCES

Aronson-Lehavi, Sharon and Rokem, Freddie (2009 [forthcoming]) 'Word and Action in Israeli Performance', in Jon Mckenzie, Heike Roms and C. J. Wan-Ling Wee (eds) *Contesting Performance: Global Genealogies of Research*, London: Palgrave Macmillan.

Fried, Michael (1985) 'Realism, Writing and Disfiguration in Thomas Eakins's "Gross Clinic", with a Postscript on Stephen Crane's *"Upturned Faces"*', *Representations* 9: 33-104.

Grotowski, Jerzy (2002 [1965]) *Towards a Poor Theatre*, ed. Eugenio Barba, New York: Routledge.

Meyer, Jerry D. (1997) 'Profane and Sacred: Religious Imagery and Prophetic Expression in Postmodern Art', *Journal of the American Academy of Religion* 65(1): 19-46.

Perez, Nissan N. (2003) *Revelation: Representations of Christ in Photography*, London: Merrill.

Sedgwick, Eve (1985) *Between Men: English Literature and Male Homosocial Desire*, New York: Columbia University Press.

• Photography - *Operation: Orfeo*: Roberto Fortuna, *jesus_c_odd_size*: Richard Sandler. *Courtesy: Hotel Pro Forma*

Congregation and Performance
Experiential Metaphysics in Hotel Pro Forma's *Operation: Orfeo* and *jesus_c_odd_size*

KIM SKJOLDAGER-NIELSEN

When the word congregation is used outside the context of the church and transferred to performance theatre, we might see the move as an old performance trick - displacement, or the disruption of the moment with a shift in understanding to follow, which aims at making apparent an otherwise possibly overlooked creative potential. New tendencies within congregation practice, theology and research into ritual can enable us to consider performance theatre as a resource for aesthetic knowledge and a creative sparring partner in the development of new forms of liturgical practice. As examples I will use the Danish performance theatre Hotel Pro Forma's two productions *Operation: Orfeo* (1993 and 2008) and *jesus_c_odd_size* (2000 and 2002), which I consider phenomenologically as forms of experiential metaphysics.

A congregation is colloquially understood as a group of people gathered in a church for service on Sunday. It is expected that they are Christians - since they are in church - and that they feel a certain attachment to the local parish. This concept is what theologians and religious researchers normally understand as a parish congregation. Parish congregation as a form of practice in Western society has recently come under severe pressure. Society has changed markedly over the last fifty years. Danish society, in particular, has moved from an agricultural society to an industrial society to an information society, and in many ways religious reform has been out of step with that change. Many religious researchers and theologians (Gibbs and Bolger; Bollmann) regard the parish congregation as a relic from a bygone era: modernism. In a parish congregation thinking is based on monocentric, rationalistic organization, linear development, text culture, local public service and consumerism (public service church). The church has on the whole neglected its own capacity for renewal by interesting itself to a too limited extent in cultural change. The shift in thinking suggested in a postmodern paradigm would suggest that the church is faced with the challenge of finding forms of practice that correspond to the conditions of contemporary life and its modes of expression. First and foremost this involves taking a position on pluralism, including globalization and competition from other religions, non-confessional spirituality, nonlinearity, a visually oriented culture and

Performance Research 13(3), pp.163-175 © Taylor & Francis Ltd 2008
DOI: 10.1080/13528160902819521

individualism. Postmodernism has resulted in two related but nonetheless different tendencies for congregational practice: (1) efforts to reshape the congregation as a unit, based on postmodernism's premises, and (2) discoveries of new forms that do not insist on participation in a community to establish the life of the church.

The first tendency expresses itself in 'emerging churches', a congregational practice Gibbs and Bolger define as:

> [C]ommunities that practice the way of Jesus within postmodern cultures. This definition encompasses the nine practices. Emerging churches (1) identify with the life of Jesus, (2) transform the secular realm, and (3) live highly communal lives. Because of these three activities, they (4) welcome the stranger, (5) serve with generosity, (6) participate as producers, (7) create as created beings, (8) lead as body and (9) take part in spiritual activities. (2006: 44-5)

The other tendency is seen in what the English theologian Pete Ward, with inspiration from the sociologist Zygmont Bauman's 'liquid modernity', has called the 'liquid church'. In contrast to the emerging church's focus on the communal sense of belonging to an active congregation and a somewhat more serene belief of the individual, the liquid church works with the idea of a *network* that will create a church with many meeting places, or communities, which to a greater degree address the seeking, non-church-goer's soul. It is a church that works directly on society's terms and that exists through 'a series of relationships and communications', which 'enable a series of flows, and these flows shape contemporary society' (Ward 2003: 43). The leading metaphor is not the familiar one that the church should be 'a ship' that floats on the water but that it should be *in* the water itself. Fixed structures are dissolved in favour of a dynamic conception of the church as something that exists between participants' activities and not through the actions of the church as an institution. In this performative approach to church activities, the liquid church resembles the emergent, as these activities often sprout up at the members' initiative or are strengthened by affiliation. Moreover, one can say that the liquid church's abandonment of the central congregational form radicalizes the emerging church's project. Where Christian mainstays have seen this liquid model as destructive, others see it as completely in agreement with living in Christ. Kester Brevin from Vaux congregation in London says:

> We see the act of the Eucharist as a powerful symbol of what we believe about Christ and the body of Christ. In the breaking of the loaf of bread, what was singular, physical and fixed in one place is split up, transformed and taken out into the cities in which we live. Church for us then is perhaps simply a network of the infected. Each time two nodes in this network communicate, church is happening, the body is evolving and Christ is being formed. So is there any commitment to one another? Of course. Otherwise the network would collapse. Is there a stress on living like Christ? Of course. It's only when Christlike activity occurs between nodes that synapses are strengthened and the body emerges.
> (Gibbs and Bolger 2006: 114-15)

CONGREGATION, THEATRE AND TRANSCENDENCE

It is through Holy Communion - which Christians see as the attracting force in a congregation - that Christ appears among the gathered through the transformation of the bread and wine. In absorbing the elements, the congregation becomes a body, *communicatio in*

sacri. This state can lead beyond the church to everyday networks by living in Christ. In the creation of a congregation, a community through Holy Communion, the church has a considerable advantage over theatre. A contemporary theatre audience is a gathering with a secular meaning: a heterogeneous group of people that have gathered at an appointed time to see theatre. Of course, with theatre sociology, one can begin to speak of various audience segments and groups that allow themselves to be drawn to specific genres or who prefer a specific theatre, but it is difficult to find one unifying factor.

Nevertheless, it is perhaps precisely such a unifying force that the twentieth century's avant-garde theatre sought to find, in a secular form, if one believes the Canadian theatre researcher Christopher Innes, who has suggested that Holy Communion can be compared with the forms of spiritual transformation that Artaud, Grotowski and others aimed at, because these performative forms would transmute 'flesh to spirit through performance' (Innes 2006: 122). In these theatrical visions and performances, especially Grotowski's, it is primarily the actor's body that is the medium through which the spiritual will be cultivated through strict discipline. There is no transcendent power that penetrates the immanence, merely a form of self-transcendence. The spectator is not active but is thought of as a witness to the performed metamorphosis, which can be said to have a similarity to the Catholic mass. The archetypal example is Ryszard Ciéslak's powerful effort in Grotowski's staging of Calderon's *The Constant Prince* (1968), which the spectators witnessed while sitting around an arena stage like an anatomical theatre. The actor played the lead character's martyrdom (the Catholic prince's imprisonment and torture at the hands of the Moorish enemy) with such great physical intensity that it was no longer simply a representation of a spiritual story but in reality became a real transformation of his bodily condition.

If the actor is transformed, what of the witness? What causes a theatre spectator's experience of involvement in a performance? Psychological factors - identification or empathy - are often what theatre builds upon. Hotel Pro Forma's performances, by contrast, deal with sense perception and the formation of meaning itself through circumventing narrative and psychology and starting instead from the material's outer physical forms. The description of spectator participation in this form of theatre demands another kind of analysis: phenomenology. We find a phenomenological understanding of what participation in theatre means via the philosophers Gabriel Marcel and Ole Fogh Kirkeby, whose work includes metaphysical concepts like 'incarnation' and 'immanent transcendence'. These phenomenologies can be coupled to what the historian of ideas Dorthe Jørgensen has called 'experiential metaphysics', thereby giving a religious aesthetic context.

The French philosopher and dramatist Gabriel Marcel's phenomenology is concerned with understanding how people participate in the world. This outward orientation presupposes an inward participation in one's own life. Marcel adapts the American philosopher William Ernest Hocking's ideas about feeling as action; feelings are cognitive and can lead to self-understanding. Joining this to his own conception of incarnation, he envisioned 'the body that feels prompted to action'. As it is recounted in J. B. L. Knox's introduction to Marcel:

> Participation ensures, according to Marcel, that I as a bodily being am not just a spectator, but fundamentally seen as a *participant*. The body, the soul and the world is - precisely because of the idea of participation - not an inextricable problem for Marcel as it is for the idealism.
>
> (Knox 2003: 24)

The idea of participation claims not only that the person is bodily situated in the world but also that a person's entire existence cannot be separated from this 'situatedness':

> To be is to participate in my existence. We can never avoid being in motion; we can never participate so little that we avoid our body or our existence. Existence and the body make it so that we participate concretely and immediately - that we always find ourselves in a situation. We are in existence as participants and not as objective spectators. (25)

It is in this radical simultaneous consideration of corporeality and existence in the concept of participation that makes Marcel's phenomenology relevant for the study of the theatre spectator. With 'participation in existence', the individual is connected to the world. Existence becomes that which is participation's inner side, while that which is outside ourselves is participation's outer side. The engagement in this other (side) or the other (one) Marcel calls 'participation in the being'. Therefore, it has to do with the intersubjective, the creating of community.

Thus a picture of phenomenology begins to emerge in which it is possible to explain what it is that makes it possible for the spectator in the proscenium theatre to experience a performance as if involved in what happens on the stage. It is the participation in existence, in which both the spectator and the fictional character/actor (or performer) are a manifestation, which makes it possible for the spectator, with the help of his own existential participation, to challenge the distance from the other and involve him/herself in this other's being; specifically, the unfamiliar observation becomes possible as a participant in the being. Furthermore, this participation does not need anything else than to sit in silence, watch and listen (Knox 2003: 57).

The difference between *participation in existence* and *participation in being* can be related to Ole Fogh Kirkeby's differentiation between limbs and body (with the event as the joining element), a phenomenology he presents in *Eventum Tantum: begivenhedens ethos (Eventum Tantum: Ethos of the Event)*. On the surface, the difference between Marcel and Kirkeby is their concept of the body. Marcel uses the body as a medium for the person's encounters and experience, while Kirkeby distinguishes the cognitive and the emotional between - respectively - the body and the limbs. The notion of the body is a primarily cognitive conception, like a spatial or virtual medium that comprises more than the limbs and lets us come to meet the space of the event. The body is an 'incorporated transcendence' that 'is *condescendence*, the zone of meaning where immanence and transcendence intersect'. The limbs are *also* the body but 'which we experience as something different from "ourselves", in one's capacity of being the event's hostage' (Kirkeby 2005: 23). We cannot relate to our body as a totality; we can only 'grasp ourselves in parts ... like *limbs*' in different ways, which Kirkeby entitles 'a pictureless "affect" (for example: the feeling of warmth on the skin or of inner pain) or like an imaginary image (the picture of sunlight, the picture of an organ), as acting and being acted

against (*poiein* and *paschein*)' (75). We never completely and entirely possess our 'body'; it is what becomes distinct in the meeting with the other aspect *that happens*. We become part of something that is bigger than ourselves, but something that is also part of us, the immanent transcendence of the event. My body meets another body in the event space, and the two zones of meaning seep into this space; the two bodies thus constitute the event. The limbs can be more or less included in the meeting, and this can be felt emotionally even if it cannot be considered physical contact. We know this aspect from the nonverbal communication's *proxemics*, that is to say, the body space or the distance between limbs (Hall 1959).

It is the event that joins the limbs and the body in Kirkeby; participation is the premise for existence and being in Marcel. To continue the comparison between the two phenomenologies, in existence the limbs are felt, while being suggests the body that ultimately cannot be delimited. Participation naturally implies both space and the event. It seems that there is nothing that immediately contradicts the two phenomenologies, which supplement each other in spite of their differences. Where Marcel's attitude to intersubjectivity through participation can seem like something effortless, Kirkeby's body/limbs distinction serves, with its elaboration of Marcel's being and existence, as a possible corrective.

Participation in being or transcendence of the self is not something we may take for granted; it is something we, with Kirkeby's understanding, must work for by approaching the event from the good side, that is to say, through an ethic, which he calls the 'ethos of heteroentitavity' (a neologism of Greek *heteros*, 'other', and Latin *entitas*, 'the entity'). It means 'to be on the Other's terms, namely the event's, because it is the event that seriously is able to reproduce our inner feeling of 'sameness' (Kirkeby 2005: 32). Only through that can we 'prove ourselves worthy of the event', as Kirkeby calls it (32).

To prove ourselves worthy of the event is, however, not an ethic in a normative sense, in the way that the ethic is understood in classic philosophy; it is rather a form of experience that makes us receptive to the incident's other, the immanent transcendent. It resembles the form that Dorthe Jørgensen finds '[a]n aesthetic, which neither is preoccupied by transforming itself to a [classical] ethic nor by focusing on the object or the subject for experience, but which on the other hand loses itself in the experience's form that is the way in to the experiential metaphysics' (Jørgensen 2003: 480; my addition). Jørgensen is preoccupied by defining how art, including installation and light art, can give an introduction to metaphysical experience. It is an 'experience of immanent transcendence', a going-beyond the normal world, which is made accessible through art language of form. The experience is not associated with a theological concept of God, but it is a secular experience of divinity, an experience of difficultly interpreted multiple meaning that says nothing about the actual existence of God but does not dismiss it either. Experiential metaphysics should create the basis for a religious aesthetic whose task it becomes to conduct 'experiential phenomenological analyses of concrete phenomenon' in proportion to the religion's resource of interpretation (Jørgensen 2002: 111–12). The opportunity for the reinterpretation of religious themes and myths emerges through this, which I think is close to what Hotel Pro Forma's staging achieves.

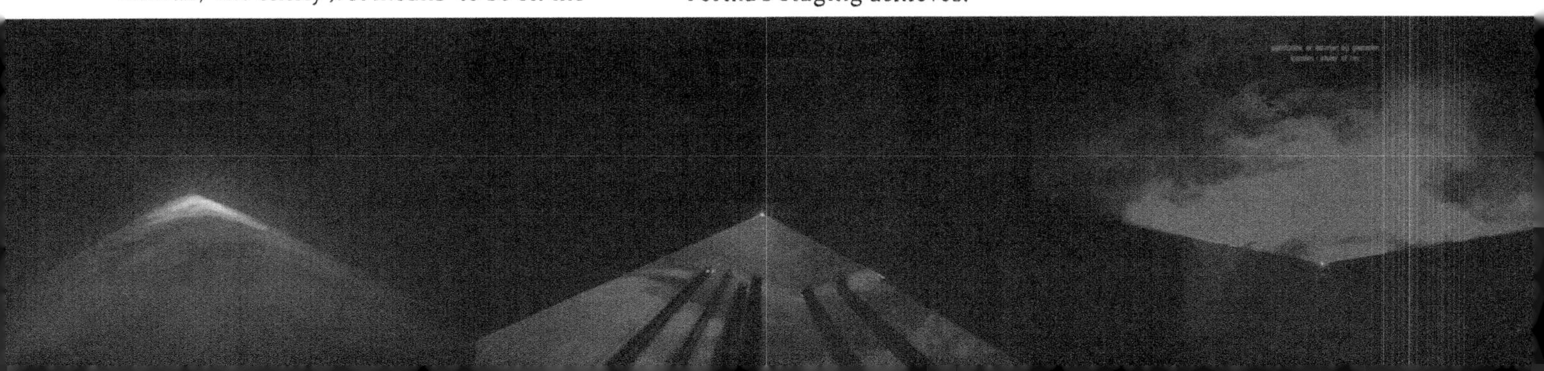

PERFORMANCE THEATRE: BETWEEN RITUAL AND THEATRE

The route to understanding performance theatre as a resource for aesthetic knowledge does not only include experiential metaphysics. It needs to include and attend to theology, as well as to research into ritual, before returning to a consideration to theatre studies and the scenic arts.

Bound as the church is to the preaching of the gospel, the production of meaning through ritual is central to practical (Protestant) theology. There has been little interest here in phenomenology. Homiletics (sermon teaching) has overshadowed the bodily and nonverbal aspects of the performance of church ritual. This disparity is being slowly remedied through the influence of recent research into ritual, as focus has shifted from text to performance, following the performative turn within the humanities. The introduction of anthropological performance theories makes the actual experience of the liturgical acts - what one does during the church service - interesting to theologians, as Bent Flemming Nielsen indicates in his book *Reenactments: Ritual, Church and Communication (Genopførelser: ritual, kirke og kommunikation).* To a greater extent what shapes meaning is active participation in ritual, rather than an understanding of the text's message, that is to say, the content of faith. But herein a problem lies for theologians: people should fully understand the message as the church proclaims it. One solution is to suggest strengthening verbal communication, which Nielsen does; but, in this scenario, churchgoers are undervalued as creative recipients who always will have (or form) their own opinions about what should transpire in the church service. To allow the churchgoer to keep his imagination to himself is an important part of the personal acquisition of faith.

In a possible further development of liturgical knowledge, and as an entry point to metaphysical forms of experience that will reach out and reflect on the church service, theatre research offers the practical theologian another view of the church service: the hitherto undervalued performative and bodily dimension of the ritual is seen in another perspective that - based on performance analysis - seeks to combine a performance's phenomenological and communicative levels. Performance theatre comes close to ritual in terms of the meaning production's openness.

Performance theatre endeavours to prevent immediate decoding by deliberately avoiding the preprogramming or pre-interpretation that we typically find in a dramatic performance. Kirsten Dehlholm, founder and artistic leader of Hotel Pro Forma, expresses it in this way:

> We are raised to suppose to understand. I would rather wait with understanding until the second after. The subconscious mind registers much more quickly than the conscious mind and I use all my performances to get below consciousness down to the pure sensation, to pre-feeling. The body recollects it, but it is very difficult to put in words.
> (Christoffersen 1998: 24)

Performance theatre seeks its ultimate in what Dehlholm calls a 'purified act', a scenic action, which can be compared with a pre-linguistic, pre-meaning *ritualization.*

In their research into ritual, the English ethnographers Caroline Humphray and James Laidlaw have defined ritualization as 'a specific departure from how things otherwise are done: it is not an intrinsic feature of all action but a particular, occasional modification of an intrinsic feature of action, namely its

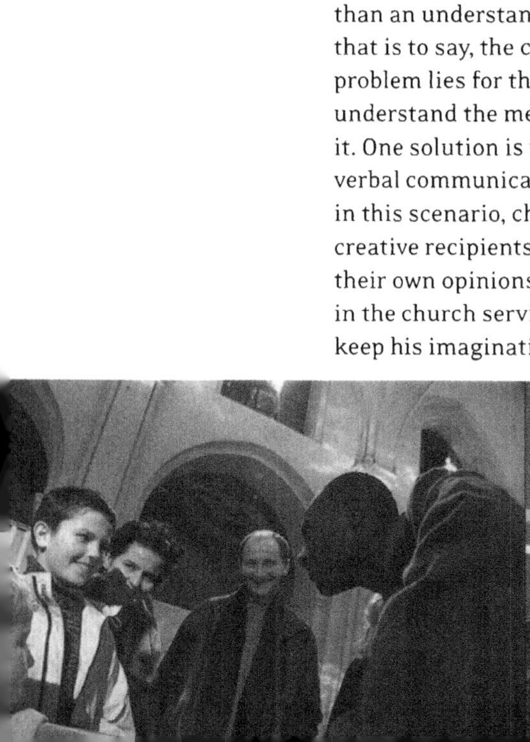
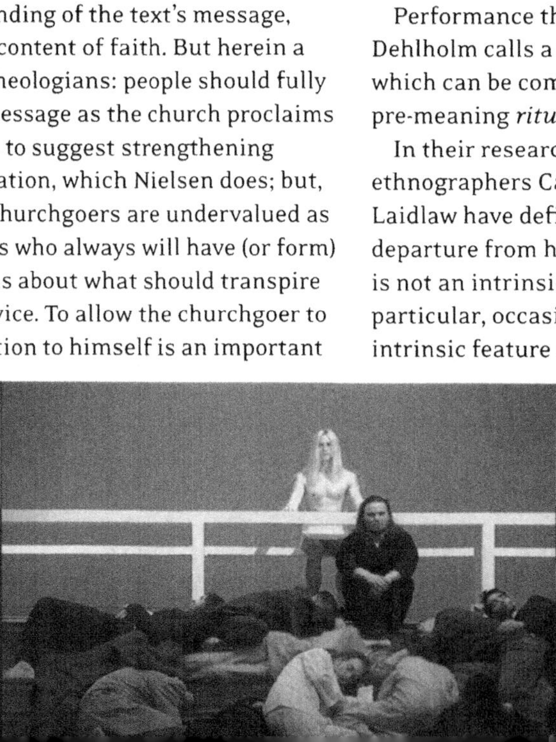

intentionality' (Humphray and Laidlaw 1994: 73). Humphray and Laidlaw take their point of departure in the sociologist Max Weber's idea for an ordinary action, which is comprised of *an act + an intention*: meaning, what we do in the everyday is always directed at a goal. For example, we eat and drink because we will satisfy our thirst or hunger. Ritualization, on the other hand, is a *distancing of the act itself from the implied intention.* We can say that the story turns into an act through altering the story's quality by having postponed its intention, which, in the same moment, empties the story and opens it to a vast number of interpretations. For example, in the Holy Communion ritual when we eat the bread and drink the wine, it does not fulfill our biological needs. A ritualized act has in and within its actual formulaic execution (or performance) no rational orientation. When the acts are put into this specific religious context, this context will however immediately intrude, inquiring into the acts' *special* significance. Here, the theological and dogmatic explanation will point to the symbolic meaning, 'We receive Jesus' body and blood with the bread and wine.' Conversely, the answer from the churchgoers could be the tautological non-answer, 'We go to Holy Communion because it is what we do' (Nielsen 2006: 55-6). Here we can see how the context (the church) closes ritualization's production of meaning, while the participants in practice keep it open. These conditions lie close to the problems surrounding performance theatre: critics and university researchers will often focus on the production of meaning in specific interpretations, while the artists will allow it to be left open and maybe therefore directly oppose interpretations.

Performance theatre's 'purified acts' have no analogy to actual rituals; the context of the performance determines the acts as ritual. The German theatre researcher Hans-Thies Lehmann utilizes a less-loaded designation about ritual-like features calling them 'ceremonies'. This brings aesthetics to the forefront:

> Postdramatic theatre ... liberates the formal, ostentatious moment of ceremony from its sole function of enhancing attention and valorizes it *for its own sake*, as an aesthetic quality, detached from all religious and cultic reference. Postdramatic theatre is the replacement of dramatic action with ceremony, with which dramatic-cultic action was once, in its beginnings, inseparately united. What is meant by ceremony as a moment of postdramatic theatre is thus the whole spectrum of movements and processes that have no referent but are presented with heightened precision; events of peculiarly formalized communality; musical-rhythmic forms or visual-architectonic constructs of development; para-ritual forms, as well as the (often deeply black) ceremony of the body and of presence; the emphatically or monumentally accentuated ostentation of the presentation.
>
> (Lehmann 2006: 69, his emphasis)

In their capacity to be framed and left to the spectator's imagination, performance theatre's acts, however, increase in value rather than remaining empty and meaningless gestures through representation as effect. The aesthetic rituals of performance theatre demonstrate ritualization in artistic form - a point that resembles the Scottish theologian William Robertson Smith's proposal (which in his time was seen as blasphemous) in his *Lectures on the Religion of the Semites*:

> So far as myths consist of explanations of ritual, their value is altogether secondary, and it may be affirmed with confidence that in almost every case the myth was derived from the ritual, and not the ritual from the myth; for the ritual was fixed and

> the myth was variable, the ritual was obligatory and faith in the myth was at the discretion of the worshipper. (Smith 1894 [1889]: 18)

Operation: Orfeo

Precisely this relationship between myth and ritual can be seen in Hotel Pro Forma's opera performance *Operation: Orfeo* (premiered at Århus Music House 1993). One can maintain that the performance, with its tightly composed visual choreography and strongly associative use of poetic text and music, plainly demonstrates Smith's assertion of ritual as unalterable and myth as interchangeable, but, as we shall see, the creative recipient (the spectator) is taken seriously, challenged by a formal interplay between the staging's meaninglessness (ritualized gesture and movement) and a known mythic material (Orpheus and Eurydice), through the yet-to-be-named (ritual acts and space) and the archetypal (light, darkness, direction - up and down), and the relationship between the immanent (the spectator subject) and the transcendent (the performance object). Meaning is left to the spectator through a form of language that simultaneously 'pulls' at ritualization and extends beyond this concept. In its phenomenological play on the spectator's senses, it almost approaches liturgically related expression: the orthodox icon's meditative form. The emergent church, creatively combining inspiration from older church ritual practice with newer (Gibbs and Bolger 2008: 179), might find *Operation: Orfeo* a contemporary artistic contribution that in its postmodern aesthetic captures old principles of spiritual emergence. This is not to say that *Operation: Orfeo* is a 'service'; no, it is an example of the actual service's premise: the experience of the metaphysical.

How does the performance, therefore, turn this relationship between myth and ritual around? It does so by not taking its point of departure in the narrative. The narrative comes in later as a frame of reference. Like most of Hotel Pro Forma's performances, it begins with a space (Dehlholm 2008: 3). The entire performance is performed as a musical-visual sequence on a large framed staircase, which constitutes its own architectonic space on the theatre stage. As such, it is a treatment of the opera from performance theatre's aesthetic perspective: first and foremost as sense perception and (emotional) movement.

As the scenic pictures are without direct references, the spectator is free to acquire the performance as a perceived metaphysical experience. The Orpheus myth is only suggested. The solitary dancer's gliding movements down, up, down on the seemingly endless stairs can, for some, bring a Eurydice figure to mind, somewhere between the living and the dead. Ib Michael's libretto, which is shown as lighted running text on a screen, can support the thematic form of loss with its sensual descriptions of diving, drowning and a burial rite. A three-tiered division within the lighting design serves to outline the myth's development. The thirteen members of the choir sing out the first twenty minutes in the dark (the descent to the land of the dead), then the next twenty minutes in a golden semidarkness (the ascent) and the final forty minutes in changing light (the loss and remembrance of the beloved). And en route the choir makes use of three distinct movements: the hand covering the eyes; the head turning to see back over the shoulder; the hand waving to mean come back. Suddenly, from among the newly composed *a cappella* chorale of Bo Holten and the avant-garde compositions of

John Cage, a soloist sings the famous aria 'Che farò senza Euridice' from Gluck's *Orfeo ed Euridice* (1762). Finally, the performers and then the spectators are drowned in a growing blue green laser wave that appears in the smoke filled auditorium as the river Hebrus on which Orpheus's head floats after he is torn to pieces by the wild maenads. In the precise, almost superhuman stylization, all scenic acts collectively contain much more than the knowledge of the myth can supply them with, namely the spectators' own emotions and associations. I experienced the production at its latest performance in Copenhagen at The Royal Theatre in 2008 as a requiem for my recently deceased mother.

The openness of meaning (or ritualization effect) increases through acts that have nothing to do with the myth: members of the choir perform small acts with yellow objects that they carry in their pockets (for example holding a piece of paper as if reading, blowing up a balloon, shaking rattles) and through various choreographic movements, which the choir performs. These small acts can create wonder or bring a smile, the choreography can create variation and fascination for the eyes, but primarily they interrupt any search for the myth and refer the spectator back to the experience of the performance itself.

This overstepping of the spectator's perception is made possible by a precise harmonization and coordination of the relation between the spectators' point of view, the stair's spatial physical dimensions, the lighting and the placement of the performers. Optical illusions emerge. In many scenic pictures the light makes the stairs appear as a two-dimensional surface with the performers seemingly painted on, for a long period merely as shadows and later like impersonal shapes. In one moment, a three dimensional space appears inside the frame, as the choir is lit from underneath with lamps hidden behind the bottom edge of the frame; by bending a little forward it appears as if they are looking down into an abyss. In another picture, by sitting diagonally on the steps the perspective is turned on its side; are the performers lying down and we see them on a slant from above? With these scenic effects, the direction and feeling of space is confused. Unnatural space is created, a space that nevertheless in the experience of the senses is real. Not only is the sense of vision affected, but all the spectator's limbs are involved in the spectacle. According to Marcel, the spectator is an existential participant in the other's being, while according to Kirkeby 'limbs' have been taken hostage by the events' immanent transcendence. When the laser wave finally breaks the frame and begins to swallow the audience, in a Kirkeby-esque sense, an 'event body' arises that encompasses all onlookers in the theatre.

This last all-encompassing effect suggests a performative parallel to the orthodox icon's 'reversed perspective'. Operating with a strongly stylized and superhuman or impersonal expression and its cancellation of the normative space - especially through the effect on the surfaces - the performance establishes a window to another, transcendent world. The icon is often interpreted, with respect to its theology, as a window to the divine eternity, but in contrast to a Renaissance painting where the perspective vanishing point is measured from the observer's privileged placement in front of the picture, the icon has no depth-of-field effect: the motif is two-dimensional and alternatively perceived as a window, through which light falls inward on the contemplator. The perspective disappears,

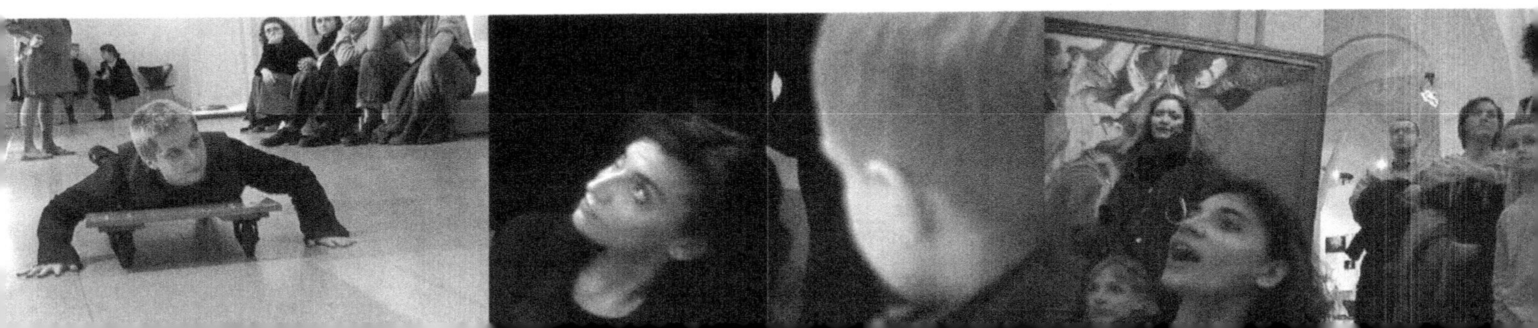

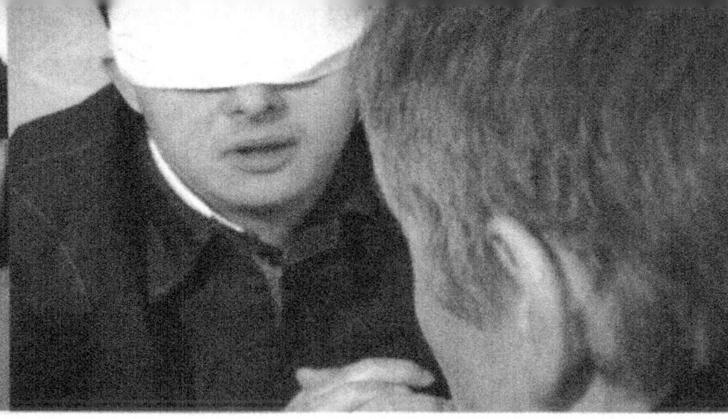

therefore, not inwardly into the picture towards a distant horizon in the motif; it turns toward the observers and embraces them instead. In *Operation: Orfeo* this 'reversed perspective' happens concretely when the scenic space spreads with the light wave, which corresponds to the theologian Ola Sigurdson's concept of

> the icon's 'third dimension', its space, which lies in front of rather than behind it, that is to say, in the real space in front of the icon. In that way the people, who are represented by the icon, are unified with the people, who observe the icon of the same capaciousness. This conforms with the icon's efforts of longing for placing the person in another world than the one to which they would ordinarily belong. (Sigurdson 2007: 227)

Seeing the theatre spectators as an emergent congregation, the analogy gives additional meaning to the orthodox icon. 'To go through with the liturgy is nothing other than to shape a living icon, just like to paint or sense an icon is a liturgical act' (275). I contend that *Operation: Orfeo*, through the realization of the metaphysical experience, has the potential for instilling in the spectator a divine view. That is the icon's purpose. However, for this effect to manifest itself in an orthodox sense, the icon must not stand alone but must be utilized as liturgy. If *Operation: Orfeo* should find application in postmodern Christian mission or practice of faith, it must be given a corresponding context. Otherwise, it is what Dorthe Jørgensen calls a profane experience of divinity, which naturally has its justification too.

jesus_c_odd_size

Operation: Orfeo's metaphysical experience of space depends on the physical distance to the spectator. *jesus_c_odd_size* works with proximity; the audience is encircled by the performance, which is like a museum exhibition, in which the performers come extremely close to the visitor. At times it becomes difficult to distinguish performer from spectator.

In relation to our discussion of the congregation and performance, *jesus_c_odd_size* will bring to mind the stuff of Christianity, the Bible stories of Jesus, not by performing these as direct illustrations but by presenting them as living, recognizable scenarios with people from our time. For example, the Jesus figure is presented as the young American cartoonist Mike Diana, who is a kind of contemporary fellow-sufferer with Jesus, because he has been sentenced to prison for blasphemy in his home state of Florida. Jesus' disciples are all young men but in theory could be anyone. Through the spectator's up-close experience of these chosen people's ritualized acts and the spectator's personal attainment, biblical pictures are transformed into 'contemporary visions' - again, the bodily, experiential approach to the transcendent, the other. In this sense, the performance views Christianity as a frame of association. As we will see, *jesus_c_odd_size* comes close to Pete Ward's liquid church form, where 'a series of relationships and communications' can arise between the visitors and the performers. And as a series of Kirkeby-esque events, the performance can create an emergent body communal for the participants - perhaps a Christ body.

In its scenic organization *jesus_c_odd_size* draws on medieval theatre's liturgical dramas, mystery and passion plays, sending the spectator wandering between various locales - spaces, tableaus, performances, conversations, installations etc. The route is not determined in advance; the scenarios play out simultaneously

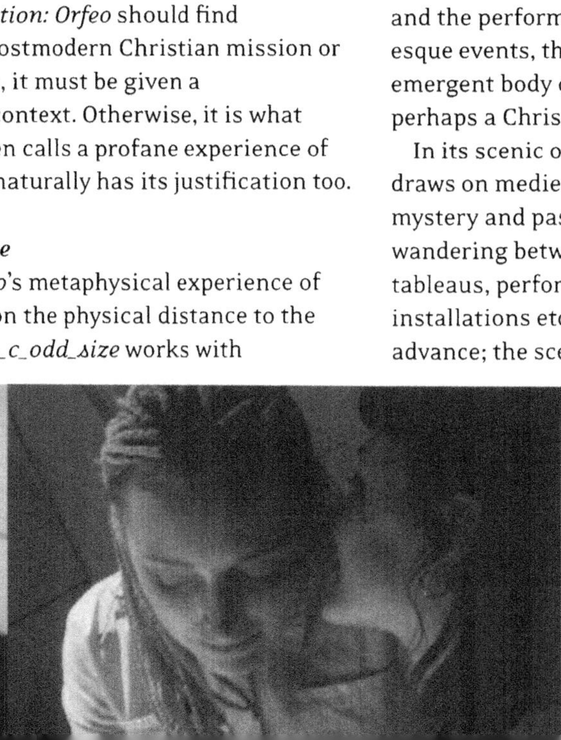
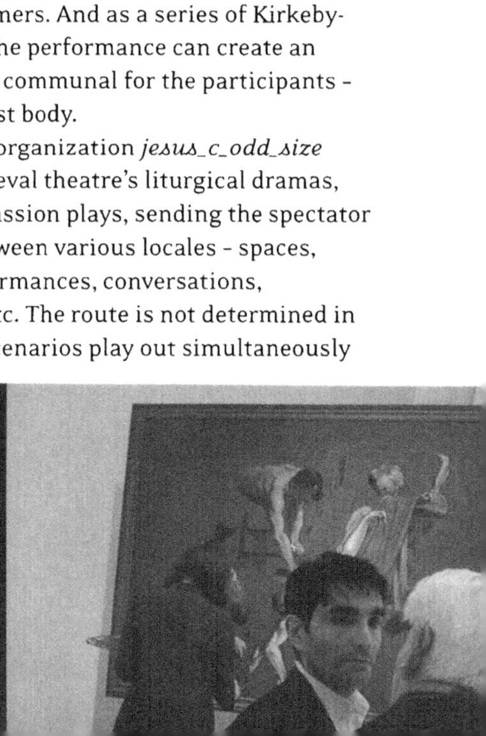

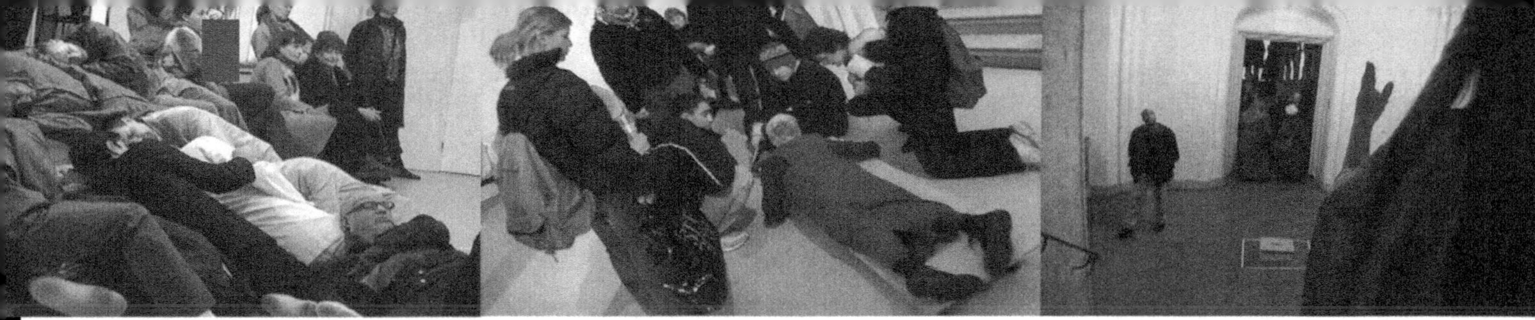

in many places, and the spectator has to make an individual choice or follow a random impulse. It is similar to visiting the Night Church in Copenhagen Cathedral, where one is not required to follow a certain liturgy but can piece together one's own church service from the optional elements.

As intimated in the way that the title is written, *jesus_c_odd_size* is an exploration of Christianity, almost as if it took place on the Internet: nonlinearly and through 'surfing'. Most critics followed this mindset and saw the performance as reflective of contemporary spirituality: a buying trip to the supermarket with free choice on all shelves. The head of the seminary in Copenhagen, Mogens Lindhardt, went another way and likened it to a vegetative stroll through a cemetery or in a cathedral, where one stops at various graves and imagines the dead. As different as they are, however, these two metaphors have one common thread: the horizontality. The audience in Marcel's concept is situated in a participation of being that offers a network of possibilities without them possessing the privileged overview of the proscenium theatre. In this way the orientation is displaced in space - from the wide angle of the overview to the snapshot of the moment. This orientation is repeated in the choice and the use of the buildings where*jesus_c_odd_size* is performed.

The world premiere took place in 2000 at a university for art, culture and communication in Malmö, Sweden. The converted industrial building's vastness and many floors accentuates the experience of wandering, where the many scenarios offer themselves en route as stations for experience and absorption. The building's somewhat ungraceful modernism plays in contrast to the biblical motifs that one meets as something simultaneously deeply unfamiliar and, nevertheless, on the surface, immediately engaging.

Two years later Jesus returns, now at a museum for contemporary art in the centre of Copenhagen. The Nikolaj Exhibition building, a former church, gathers the performance in its nave, maybe consolidating the feeling that in this former church space Jesus has 'come home'. Again, the entire building is used, the side aisles, tower chambers, the galleries and lofts, and as a visitor one need not take many steps before one literally walks into the next scenario.

The staging consciously plays on the proxemics of the body's space and nonverbal communication. The Last Supper, which is staged on the church's apse on an existing platform, is intended to be seen from a distance, in the public zone (3.5m - 7.5m). Others, like Golgotha (Calvary), can be experienced from a distance or in close proximity (the personal zone: 45cm - 120cm). Freedom of choice in distance applies especially to Golgotha, because here the human is exhibited in all its dreaded vulnerability: the performers in the posture of the cross hang in large vacuum packed plastic bags in front of the spectators. It raises the issue of contemporary man's repression of suffering and death in plastic, opposite Christianity's acceptance.

Both randomness and coincidence are important factors in the beginning of the meetings, which is most clearly seen in the personal zone or intimate zone (15cm - 45cm). It is probably the situation where the experience of immanent transcendence, of over-indulgence of self, is the strongest with the audience. The experience can arise in exchanges between the associative biblical frame and the audience's active, bodily investment in the various performances.

In the loft over one of the church's side aisles, an old-fashioned furnished living room is found in which a gathering of older women is holding a coffee klatch. The coziness is contagious. They are the disciples' grandmothers. On the two tables stand pictures of their grand children and they speak most lovingly and admiringly of them to the audience. One might be offered a place at the table, where cake and coffee is served. The situation is doubly curious because here one speaks with an ordinary, full-of-life person, who one concurrently sees as a freely biblical fabrication. One is woven into the performance, which there and then has become interactive.

In two tower chambers, the Virgin Mary receives visitors. Here she sits at a table and invites the visitor into conversation, enveloped in a material red neon light as if it were in a chamber of the heart. Now to a higher degree than before, we step over a threshold and into another room, where the whole sensual materiality will blend with our Kirkeby-esque body space. It is the fascination of the space that lures one inside into this encounter's event. In a literal sense it is a strange, theatrical space that belongs to the performance about the Virgin Mary. One expects a portrayal of the Virgin Mary, but she is yet another ordinary person, a girl named Helle, conversing about what now must transpire. The disappointment can be great for some, if they do not adapt to accepting the event's premises. But within the disappointment also lies a realization that the Virgin Mary, no matter how removed and raised she ended up being in the Catholic church, still was once just an ordinary girl with whom one could have a trivial conversation.

Back at the church's nave one can also encounter Mary Magdalene, who offers to wash the visitor's feet just as in the legend when she washed Jesus' feet. Suddenly one has a choice to say yes or no. Foot-washing is an intimate and private phenomenon. The performer is obliging and smiling and tries in the most respectful way to signal confidence. With Kirkeby, we can say that it involves not only one's body as a zone of meaning but particular limbs. On the plane of meaning, one becomes even challenged to replace Jesus in the picture. Does one want to do that? Or does one perceive her through the contemporary problematic of gender as a submissive person who one does not wish to oppress further? Does the guest understand the devotion that from the Christian side is attributed to the ritual?

When one is either deeply touched or modestly reluctant upon leaving the stage, there is a good chance that one will happen on a leper. Two paraplegic girls ride around sitting and lying on boards with wheels in the nave's transept. Like real lepers, they have ringing bells that warn of their arrival. Upon a sudden meeting in the personal zone, one can only awkwardly excuse oneself. The girls try to make eye contact, while they sing the most beautiful song. This is disquieting until we, assured by the girls' genuine proximity, can be fully accepting. Both parties are exposed in the meeting, but by insisting on it - showing ourselves worthy of the event, as Kirkeby would say - an event's space arises where all bodily disparities and flaws seem unimportant: a sameness between the parties is established. Such a meeting brings us, as visitors, to exceed ourselves and our potential prejudice. A charitable look can arise; one is equal, face to face.

'We always meet the Other on the threshold. A hesitation, it is the picture of the meeting with the Other in the event' (Kirkeby 2005: 210). Kirkeby sees the challenge in between as the humane meeting. Should I as a non-theologian

venture a last assertion, it would be that even Christianity as practice always revolves around coming over the phenomenological threshold, as Jesus crossed over it and with that into another person's life. Whether *jesus_c_odd_size* is able to create a liquid church, as Pete Ward understands it, I will not ultimately say, but I will suggest that the performance could be seen as an exercise in that direction.

Translated from Danish by Whitney Byrn.

REFERENCES

Bollmann, Kaj (2003) 'Mellem sogn og cyperspace' (Between Parish and Cyberspace), in Morten Thomsen Højsgaard (ed.) *Den digitale kirke: Syv artikler om internet og kristendom* (The Digital Church: Seven Articles on Internet and Christianity), Frederiksberg: Anis, pp. 41–53.

Christoffersen, Erik Exe (1998) *Hotel Pro Forma*, Århus: Klim.

Dehlholm, Kirsten (2008) *Hotel Pro Forma: Operation: Orfeo*, København: Skoletjenesten (teaching material).

Gibbs, Eddie and Bolger, Ryan K. (2006) *Emerging Churches*, London: Ashford Colour Press.

Hall, Edward (1959) *The Silent Language*, New York: Anchor Books.

Humphray, Caroline and Laidlaw, James (1994) *The Archetypical Actions of Ritual: A Theory of Ritual Illustrated by the Jain Rite of Worship*, Oxford: Clarendon Press.

Innes, Christopher (2006) 'Artaud - Grotowski - Stelarc: Transmuting Flesh to Spirit through Performance', in Bent Holm (ed.) *Tro på teatret* (Faith in Theatre), Frederiksberg: Multivers, pp. 122–44.

Jørgensen, Dorthe (2002) 'Guddommelighedserfaring I en moderne verden' (Divine Experience in a Modern World), in Niels Grønkjær and Henrik Brandt-Pedersen (eds) *Interesse for Gud* (Interest in God), Frederiksberg: Anis.

Jørgensen, Dorthe (2003) *Skønhedens metamorfose. De æstetiske idéers historie* (The Metamorphosis of Beauty: The History of Aesthetic Ideas), Gylling: Syddansk Universitetsforlag.

Kirkeby, Ole Fogh (2005) *Eventum Tantum: begivenhedens ethos* (Eventum Tantum: The Ethos of the Event), Frederiksberg: Samfundsfag.

Knox, J. B. L. (2003) *Gabriel Marcel: Fortvivlelsens dramatiker og håbets filosof* (Gabriel Marcel: Dramatist of Dispair and Philosopher of Hope), Gylling: Syddansk Universitetsforlag.

Lehmann, Hans-Thies (2006) *Postdramatic Theatre*, London and New York: Routledge.

Nielsen, Bent Flemming (2004) *Genopførelser: ritual, kirke og kommunikation* (Reenactments: Ritual, Church and Communication), Frederiksberg: Anis.

Nielsen, Bent Flemming (2006) 'Troens gestik: relationerne mellem ritual og teater' (Gestures of Faith: Relations between Ritual and Theatre), in Bent Holm (ed.) *Tro på teatret* (Faith in Theatre), Frederiksberg: Multivers, pp. 48–72.

Sigurdson, Ola (2006) *Himmelska Kroppar: Inkarnation, blick, kropslighet* (Heavenly Bodies: Incanation, Gaze, Embodiment), Munkedal: Glänta Produktion.

Smith, William Robertson (1894 [1889]) *Lectures on the Religion of the Semites*, London: Adam and Charles Black.

Ward, Pete (2003) *Liquid Church*, Carlisle: Paternoster Press.

The Drama of Liturgy and the Liturgy of Drama

SAMUEL WELLS

INTRODUCTION

I begin with an ironic reversal. Liturgy was originally a public act of benefaction. Drama was originally a religious ritual. Today liturgy broadly means a religious ritual, while drama is more of a public event. Let me explain.

The origins of liturgy lie in ancient Greece. In the classical Greek city-states the role of the wealthier citizens was to perform public duties (i.e., leitourgia). In Athens, for example, such offices included providing a trireme, or battleship (the role of the trierarchus), providing a banquet (the role of the hesitator) and providing sporting facilities (the role of the gymnasiarch). Thus λειτουργια (leitourgia) was a public work arranged by a wealthy citizen for the good of the people.

By contrast the origins of drama - in India, Japan and China as much as in Europe - lie in religious ritual. The roots of drama again lie in ancient Greece, with the recounting of stories of heroes. These tales were told by a company of voices - the chorus. As stories became more complex and more vivid, the leader and other chorus members began to take specific parts. Dialogue between characters gave way to interactions beyond the merely verbal, and drama was born. A broadly similar process took place in medieval Europe. Mystery plays and miracle plays began life on the steps of major church buildings on festival days, offering vernacular spectacle to parallel the Latin liturgy taking place inside. Such dramas later took on a life of their own, with travelling players moving through the countryside from one town to another, rivalling the official liturgy much as the mendicant friars rivalled the regular priestly hierarchy.

In the spirit of a truly international conference, I wish to step across four continents to argue that a faithful Christian liturgy must be truly dramatic. This means that liturgy needs to recognize and celebrate its origins in drama and that the Christian churches need to recognize and celebrate the drama of liturgy. Liturgy is, in short, religious drama - a public performance of a sacred story.

A HEGELIAN CONTRIBUTION

Imagine a notable person, one who attracted many followers, performed memorable deeds and offered awesome teachings, actions and words that invited projections and longings to cluster around his identity and programme. Imagine he was frequently to be found in dispute with those in authority, argument over his teachings and theirs, controversy over his actions and theirs, antagonism that threatened a violent confrontation. And imagine that when that confrontation came it was short and brutal, resulting in the man's rapid execution and the scattering of his followers. And yet that his followers regrouped, spoke of his reappearance, of their empowerment, of a remarkable force urging them to set out his programme as the culture of a new society and the anticipation of a permanent future.

Such a story might be told in a number of ways.

Performance Research 13(3), pp.176-183 © Taylor & Francis Ltd 2008
DOI: 10.1080/13528160902819539

The philosopher G. W. F. Hegel speaks of three broad ways of exploring a story in his work on poetics (1979: 142-200). The first approach he calls 'epic'. An epic story is told on the broadest possible canvas. It seeks to give the widest possible significance to the tale it tells. Its concern is not so much the particular individuals at the centre of the story as humanity in general; not so much the particular dilemmas and decisions of the protagonists as the human condition as such. Thus in the story of the notable person just related, an epic rendering would stress the general features of the narrative. It might be retold as a story in which love comes into the world, to which the world could only respond with admiration, hatred or fear. It might continue as a tale of how love and power could not coexist without power seeking to destroy love, yet a revelation of how love always wins out in the end. It could even be told as a story in which love and power inevitably clash until love lays down its life at the foot of power, whereupon love subverts power until (in the empowerment of the followers) power and love are harmoniously reunited. Thus in more or less general terms the particular features of the narrative are downplayed and the narrator steers away from a personal investment in the events and outcome.

The 'lyric' approach is significantly different. If the epic is the account from 'outside', the self-consciously objective viewpoint, then the lyric is the account from 'inside', the subjective viewpoint. Lyric speaks from the heart. It explores the depth of personal commitment and feeling and the spectrum of human qualities and perceptions involved in the narrative. Thus in the story of the notable person, a lyric account would focus on the yearning to share a precious gift, the bewilderment of misunderstanding and the agony of betrayal. It would dwell especially on the physical and mental anguish of execution, on the misgivings and deliberations of those enforcing vindictive power, on the followers' sense of loss and despair, on their incipient and sudden discovery of reversal, on their profound experience of empowerment, on their eager longing for final fulfilment. Not only the narrator but also the listener or spectator is drawn into such a vivid experience of the events that they seem as if they are indeed happening now: the boundaries between history and the present, factuality and plausibility, event and interpretation are blurred, and this story, however particular, becomes the centre of attention by the sheer force of its own telling.

The third approach is called the 'dramatic' and, this being Hegel, it is characteristically a synthesis of the first two. A truly dramatic work does not just combine action and words, it brings together the intensity of the lyric with the scope of the epic. It is about heart and head, general and particular, the human condition and the individual life, passionate experience and time-honoured wisdom. As to the story of the notable person, a truly dramatic account would seek to display both the particularity of the historical circumstances and the way those circumstances illuminated and transformed all other circumstances. Whereas an epic approach might see this story as one that simply illustrated eternal truths that were already established, a dramatic approach brings a genuine identification of how the particular transforms the general. This man's words were in one context, yet they speak to all; his actions were in a limited sphere, yet they have cosmic significance; his disputes were circumstantial, and yet they characterize every conflict; his death was obscure, and yet it can still be described as the end of much more than the life of one person; his followers' empowerment was hard to describe, and yet it can come to define the whole meaning of power.

Here we have a highly promising way of exploring what it might mean to speak of 'good' liturgy. Good liturgy should be truly dramatic. The language (of epic and lyric) is also helpful in naming and identifying some of the pitfalls liturgy can fall into. For example, the lyric style of spirituality names exactly the kind of religious expression that focuses almost exclusively on the self and its emotional dimensions and relations. The language of spirituality is notoriously amenable to whatever interpretation an individual may choose to place upon it. It has relatively little place for accountability, tradition or connection to the lives of those significantly different from itself. Truth becomes an internal, subjective matter with little purchase on public events or historical change. By contrast the epic form of liturgy can become obsessed by outward forms, historic traditions, written documents and cyclic patterns. It can become a slave to civic purposes and public expectations. Here there is dignity, formality, reserve, discipline and order: but having so carefully eradicated any trace of lyric sensibility, epic liturgy can simply lack heart - a human touch. Good liturgy has the immediacy and stirring quality of the lyric together with the breadth and authority of the epic.

AN AFRICAN CONTRIBUTION

Vincent Donovan describes in his book *Christianity Rediscovered* the profound ways in which the Masai tribe of Tanzania, to whom he ministered as a missionary Roman Catholic priest in the 1960s and 1970s, already embodied significant aspects or analogies of Christian theology and practice. Donovan offers several compelling examples of what I am calling dramatic liturgy, of which I shall here draw on three.

> If a son offended his father seriously, this was considered a sin of great magnitude ... The son was banished from the community ... Sometimes the peers of the father would encourage him to seek the 'spittle of forgiveness' so that he could forgive his son and bring blessing once again on the village. Spittle was not just a sign of forgiveness. It was forgiveness ... I [once] sat with [an old man] in the middle of the night as he prayed in vain for the spittle of forgiveness.
>
> If word does come that the spittle of forgiveness has been granted his father, [the son] will be earnestly entreated by his peers to take advantage of it. They will accompany him back to the village. And his father will be waiting with the other elders. The two groups will cross from different sides of the village towards each other in the centre. When they arrive there together, the son will ask his father's forgiveness, and the father will spit on him, and forgiveness comes, and there is great rejoicing.
>
> (1982: 59-60)

Here is truly dramatic liturgy. The friends of the father and the son plead with them to restore the well-being of the village by coming out of their lyric enmity. Yet this is no epic liturgy, no bland proclamation of forgiveness and reconciliation in the absence of personal cost and genuine restoration. The liturgy is an intensely personal one, but one in which many members of the village participate. It is a genuinely transforming practice.

An even broader liturgical dimension is portrayed in another account of forgiveness, this time on a corporate level concerning families or other social groups. Such discord is disastrous for a nomadic people if they are to preserve their herd and be safe from enemies.

> If at all possible, both the offending and the offended family must be brought back together by an act of forgiveness sought and bestowed. So at the behest of the total community both families prepare food ... This holy food is brought to the centre of the village by the two families accompanied by the rest of the community, encouraging both families all along the way. There in the centre of the village the food is exchanged between the two families, each family accepting the food prepared by the other family. Then the holy food is eaten by both families, and when it is, forgiveness comes, and the people say that a new [covenant] has begun. (60-1)

Once again, this is a truly dramatic liturgy. Not allowing the families to get lost in their own intractable lyric sensibilities, their friends bring them to an epic awareness of the significance of their disunity for the good of the whole people. As Donovan notes, 'a new testament of forgiveness is brought about by the exchange of holy food' (61). This is the heart of the Roman Catholic understanding of liturgy, and Donovan found it already being practiced by the Masai of Tanzania.

A third example refers not to a personal relationship, nor a corporate one, but to the whole community. When Donovan arrived in a community to join in worship, he would stoop down and pick up a tuft of grass and give it to the first elders who greeted him. Grass was the heart of Masai life, since their cattle lived off it.

> During stormy and angry arguments that might arise in their lives, a tuft of grass, offered by one Masai and accepted by the second, was an assurance that no violence would erupt because of the differences and arguments ...
>
> The leaders did decide occasionally that, despite the prayers and readings and discussions, if the grass had stopped, if someone, or some group, in the village had refused to accept the grass as the sign of the peace of Christ, there would be no Eucharist at this time. (124-5, 127)

Again this is a liturgy that is not just epic - such a liturgy would have gone ahead, whether there was true reconciliation or not. Neither is it just lyric - for it does not address simply the dissent among particular individuals but addresses the good of the whole community, placing the powerful feelings of the antagonists in a much larger context. Such a liturgy is truly dramatic, for it both portrays and enacts the way the reconciliation of the Eucharist is embedded in and to some extent dependent on the active participation of all members of the community.

A SOUTH AMERICAN CONTRIBUTION

William Cavanaugh's book *Torture and Eucharist* describes the response of the Roman Catholic church in Chile to the regime of General Augusto Pinochet from 1973-90. Cavanaugh begins by arguing that torture, as practiced by Pinochet's regime on a great number of its people, is a form of liturgy. He says: 'Torture is a form of perverted liturgy, a ritual act which organizes bodies in a society into a collective performance, not of true community, but of an atomized aggregate of mutually suspicious individuals' (Cavanaugh 1998: 12). He goes on to say that 'Torture is not a merely physical assault on bodies but a formation of a social imagination', seeing social imagination as 'that vision which organizes the members into a set of coherent performances, and which is constantly reconstructed by those performances' (12). In summary, torture was 'a central rite in the liturgy by which the state manifested its power' (12).

Cavanaugh's argument brings together the two themes with which I began this essay. First of all, by describing the oppressive actions of the Chilean state as a liturgy and torture as 'an assault on social bodies', he echoes my opening observation that liturgy began life in Ancient Greece as a public work. To use Cavanaugh's words, liturgy was originally and remains the construction and depiction of a social imagination - in the case of Pinochet's Chile, an imagination based around fear and complete submission to the state. Second, Cavanaugh's account of the liturgy of torture probes further into my use of Hegel's distinctions between epic, lyric and dramatic as I have applied them to liturgy. What Cavanaugh's argument means, when translated into the argument of this essay, is that torture was not just a lyric expression of a violent search for information and form of punishment: it was also an epic portrayal of the realities of the relationship between the state and its people. Because torture was not just lyric but also epic, it was indeed dramatic, and that is why it was such a profound and damaging threat to the Catholic Church in Chile. Pinochet's liturgy was not just a ghastly parody but a rival claim to truth.

Torture is historically associated with a search for information: this was often the practice in medieval Europe. However what Cavanaugh discovered was that the agents carrying out the torture were invariably already in full possession of any significant information their victim had. Not only are there countless examples of a prisoner finally giving up information in the face of extreme brutality only to be told by the torturer, 'We already knew'; but meanwhile there were at least as many confessions that both tortured and torturer knew to be false (28-9).

Why then the drama of shouted questions, inflicted pain and feigned urgency? Cavanaugh suggests that 'by making the seeking of important answers seem like the motive for the torture, the torturer seems able to justify his brutality' (29). Simply inflicting merciless pain on a helpless victim is indefensible until one brings in a sense of an urgent search for information, whereupon the process begins to seem justifiable and thus imaginable. This is how Cavanaugh summarizes the enactment of torture as liturgy:

> The victims are made to speak the words of the regime, to replace their own reality with that of the state, to double the voice of the state. The state's omnipotence becomes manifest in the horrifying production of power, what [Elaine] Scarry calls 'a grotesque piece of compensatory drama'. Torture may be considered a kind of perverse liturgy, for in torture the body of the victims is the ritual site where the state's power is manifest in its most awesome form. Torture is liturgy - or, perhaps better said, 'anti-liturgy' - because it involves bodies and bodily movements in an enacted drama which both makes real the power of the state and constitutes an act of worship of that mysterious power. (30)

In conclusion Cavanaugh says torture is 'the ritual site at which the state produces the reality in which its pretensions to omnipotence consist' (30-1).

The response to state torture in Chile could not be simply lyric. It had to be truly dramatic. Early forms of resistance to torture were limited to defiant words and hopeless courtroom battles. A new form of liturgical resistance came about with the Sebastian Acevedo Movement against Torture, named after a construction worker who died in despair in front of Concepción cathedral after searching in vain for three days for his abducted children. The movement conducted what Cavanaugh calls 'public ritual acts of solidarity and denunciation that members would perform with their bodies' (274). The actions lasted no more than a few minutes and were generally halted when police appeared and the members of the movement rapidly dispersed. The veil of fear was drawn back for a few moments, and the regime was exposed for what it was.

> In an astonishing ritual transformation, clandestine torture centres are revealed to the passers-by for what they are, as if a veil covering the building were abruptly taken away. The complicity of other sectors of the government and society is laid bare for all to see. The entire torture system appears on a city street. Techniques of torture are detailed, places of torture identified, names of victims and names of those responsible - including sometimes the names of the immediate torturers themselves - are made publicly known. Victims are thus transformed into martyrs, as their names are spoken as a public witness against the powers of death. (275)

Cavanaugh demonstrates that the liturgy of the Sebastian Acevedo Movement was what I am calling truly dramatic when he concludes that '[t]he repressive apparatus is made visible on the very bodies of the protesters as they are beaten, tear gassed, hosed down and dragged away to prison ... The ritual is designed to make the tortured body, which has been disappeared by the state, miraculously appear in the bodies of the protestors' (276). The sharing of pain in this liturgy directly addresses the central power of torture, which is to isolate the victim through the imprisonment of pain.

Thus Cavanaugh presents us with two rival forms of liturgy, each of which seeks to portray a public truth. Each seeks to be dramatic, in that each demonstrates not just the lyric expression of intense feeling but the epic portrayal of outward reality. A significant difference between them is that the Sebastian Acevedo Movement performed its liturgies in public while the Pinochet regime conducted its perverse liturgies in private. These liturgies and Cavanaugh's analysis advance my argument that truly dramatic liturgy is a public work embodying and creating a depiction of personal and social truth.

A EUROPEAN CONTRIBUTION

I have portrayed liturgy as the 'dramatic' embodiment of both personal ('lyric') expression and the public ('epic') portrayal of reality. I have traced this synthesis through an African tribe and a South American regime and its dissidents. For a third example I turn to the medieval European tradition of carnival, particularly as portrayed by the American social scientist James C. Scott.

Scott points out that carnival is a general feature not only of European culture but of a wide range of societies.

> There are scores of festivals, fairs and ritual occasions that share many of the essential features of carnival itself. The Feast of Fools, charivari, coronations, periodic market fairs, harvest celebrations, spring fertility rights and even traditional elections share something of the carnivalesque. Furthermore, it is difficult to find any culture that does not have something on the order of a carnival event in its ritual calendar. Thus there is the Feast of Krishna (Holi) in Hindu society, the water festival in much of mainland Southeast Asia, the Saturnalia in ancient Roman society, and so on. (1990: 173)

While carnival does not necessarily have a purchase on the transcendent, it is significant for my argument because again it brings together lyric and epic elements in a dramatic whole.

Carnival is one moment in the year to name and set right those relationships in which power has been abused and the vulnerable have been trodden down. It provides an anonymous setting for the proclaiming of accusations that would otherwise be confined to gossip, and for the ridiculing of those usually too puffed up to be subject to criticism. It delights and exults in bodily functions - eating, drinking, defecating, fornicating, burping and farting - because such activity brings all people to an animal level of common, earthy humanity where no one can be superior. For servants in grand houses, great institutions, places of worship and in the homes of the well-to-do, whose lives were circumscribed by ritual and restraint, carnival was a release of tension, an articulation of vernacular idiom, a window of honest discourse amid a lifetime of subservience and pretence.

> The young can scold the old, women can ridicule men, cuckolded or henpecked husbands may be openly mocked, the bad-tempered and stingy can be satirized, muted personal vendettas and factional strife can be expressed. Disapproval that would be dangerous or socially costly to vent at other times is sanctioned during carnival. It is the time and place to settle, verbally at least, personal and social scores ...
>
> Much of the social aggression within carnival is directed at dominant power figures, if for no other reason than the fact that such figures are, by virtue of their power, virtually immune from open criticism at other times. Any local notables who had incurred popular wrath - merciless usurers, soldiers who were abusive, corrupt local officials, priests who were avid or lascivious - might find themselves the target of a concerted carnival attack by their erstwhile inferiors. Satirical verses might be chanted in front of their houses, they might be burned in effigy, and they might be extorted by masked and threatening crowds to distribute money or drink and made to publicly repent. (173-4)

Scott's treatment alludes to the contrasting understandings of carnival held by scholars in the field. The conventional view is that carnival is a deliberate method employed by social elites to extract the stopper from the bottle from time to time lest the bottle otherwise explode from the energy inside it. Thus the tradition of Mardi Gras is to cram into one day all the physical and sensual longing and desire that will be once more pent up when Lent begins the following day. It is argued that social elites orchestrate a day of misrule in order subtly to underwrite the power relations that exist the rest of the time. Scott counters this prevailing view by pointing out how strongly elites work to neuter or prevent the activities of carnival. If carnival were genuinely established as a safety valve, one would expect to see those controlling social speech and space actively encouraging carnival - but there is little sign of this. Scott notes that banning carnival was one of the first measures brought forward by Francisco Franco's Nationalist government on taking power in Spain in 1936. Scott recognizes the way carnival releases social tension, but is more inclined to see carnival as 'an ambiguous political victory wrested from elites by subordinate groups' (178). Carnival is not just a ritual modelling for revolt but can be an actual dress rehearsal or even a natural setting for rebellion.

Such a portrayal of carnival adds a third ingredient to the dimensions of dramatic liturgy offered by Donovan's account of the Masai and Cavanaugh's account of Chile. In Donovan's account we saw how dramatic liturgy involves tangible elements such as spittle, tufts of grass and shared meals. We saw how such liturgy weaves together the intensity of personal animosity or estrangement with the communal need for reconciliation and forgiveness. And we saw how such liturgy not only creates a society but displays it, modelling the forms of life that sustain and renew the community. In Cavanaugh's account we saw even more strongly the physical nature of liturgy. By recognizing the way in which torture is a liturgy that dismembers social bodies, it became clearer how dramatic liturgy is a positive process of re-membering - making present those who have (been) disappeared and weaving together stories of personal anguish and social oppression. We saw how the public performance of litanies that named the truth of what was taking place made it possible for the social imagination of the country to comprehend the evil that was taking place in secret. Now in Scott's account of carnival we can see the confluence of the lyric settling of personal scores and indulgence of bodily functions together with the epic battle over who tells the true story of society and whether that story is one of order and conformity or subversion and revolt.

CONCLUSION

The distinction between epic, lyric and dramatic liturgy offers a language with which not only to name some of the false dualisms in liturgy but also to identify what might make some liturgies better than others.

Truly dramatic liturgy must always challenge a number of false dualisms, of which I mention three by way of illustration. There is, first of all, the dualism between the mental and the physical. This is not inherently a distinction between epic and lyric, although the distinction is helpful in drawing out the kinds of gestures and words that may contribute to good liturgy. The examples offered by Donovan, Cavanaugh and Scott show how reconciliation, protest and subversion require not just verbal litanies but embodied practices if they are to create and display the social imagination of a people. But these examples also show that neither drama nor liturgy is ever purely verbal or purely physical. Protesting against torture or speaking through a mask at a carnival is never just a verbal matter, but an embodied liturgy.

A second false dualism is between the personal and the public. A lyric liturgy will be so eager to stress the poignant, the passionate and the pressing that it will invariably risk sentimentality, self-deception and self-centredness. An epic liturgy will be so anxious to be dispassionate and impartial and respectful of established norms that it will invariably risk losing a human touch, underwriting the status quo and offering a transcendent justification for existing oppression. Truly dramatic liturgy challenges what it sees as a false distinction between the personal and the public. As all the examples in this essay have shown, no such line can ultimately be drawn. A dramatic liturgy so construes the personal that it is drawn into and narrated through the greater story of God, and so construes the public that it is at once perceived through the prism of personal commitment and divine providence.

A final false dualism is between affirmation and subversion. Simply to name this as a dualism in liturgy is to recognize that in many liturgies in apparently stable societies the dimension of protest, let alone subversion, seems to be almost wholly absent. And yet at the centre of the Christian faith is the execution by an extraordinarily painful method of a man who was undoubtedly seen as a threat to the political, social and religious elite of his day. The connection between the examples of Donovan, Cavanaugh and Scott is this: the formation of the kind of community with the kind of liturgical and social practices described by Donovan will almost inevitably make such a community a threat to power elites (as in Pinochet's Chile), and this in turn will require an imagination shaped by liturgy to offer appropriate kinds of protest or resort to the parody and satire of carnival in order to witness to the kind of society the Christian faith proclaims. For example, when a Christian priest in liturgical robes walks up the aisle of a church at the beginning the liturgy, is this practice perceived as the celebration of a royal priesthood, or is it considered a re-enactment of Christ's walk down the via dolorosa to the cross? If it has royal connotations, are these ones that reinforce models of kingly authority in the wider society, or is it a defiant display of the recognition of a very different king? If it has connotations of Christ's walk to the cross, is this in subversion of the pomp and pride of the wider society and perhaps even the gathered congregation itself, or is it in solidarity with the oppressed among the congregation or elsewhere? These are exactly the questions that should be asked about any practice of liturgy.

This paper was read at the First Tehran International Conference on Drama and Religion, 7-8 January 2007.

REFERENCES

Cavanaugh, William T. (1998) *Torture and Eucharist: Theology, Politics and the Body of Christ*, Oxford and Malden, Massachusetts: Blackwell.

Vincent J. Donovan (1978) *Christianity Rediscovered: An Epistle from the Masai*, London: SCM.

Hegel, G. W. F. (1979) *Hegel on the Arts: Selections from G. W. F. Hegel's* Aesthetics *or* The Philosophy of Fine Art, abridged and trans. Henry Paolucci, New York: Frederick Ungar.

Scarry, Elaine (1985) *The Body in Pain: The Making and Unmaking of the World*, New York: Oxford University Press.

James C. Scott (1990) *Domination and the Arts of Resistance: Hidden Transcripts*, New Haven: Yale University Press.

Mouth to Mouth, Body and Blood
Self in congregation

EMILY ANDERSON

When my dad lifts his arms in their wide white sleeves, everyone stands up; we are, each one, one of his arms, moving with the impulses moving through his cerebellum and spine beneath alb and stole, suit and coat and collar, cincture and skin.

He says, 'All rise and prepare to commune.' A ghost darts from synapse to synapse and each of our bodies becomes one of his arms, two hundred arms, both of my arms his arm and half of me comes from my mom. We stand ready. He pours the wine from silver carafe to silver chalice, we Lutherans in Wisconsin believe in something between transubstantiation and mere symbolism, the bread and wine are more than symbols but less than a body.

Something in between or neither.

> There is an ancient concept for community without uniformity, and personality without individualism; it is the term *perichoresis* ... the noun means 'whirl or rotation', the verb means 'going from one to another, walking around, handing around (for example, a bottle of wine ...) encircling, embracing or enclosing'. (Moltmann 2000: 113)

I make my arms an O. I make my mouth an O. Say 'Oh.' Make your arms my arms, make your mouth like mine.
Turn in a circle. Say a-lay-loo.

Put your lips to the lip of the cup that fits upended on the pinkie or ring fingers, that fits also in finger-holes in the silver wheel cradled in the acolyte's arms. The church is a hole and when my dad goes - those winter nights of long council meetings - mom flips pancakes from the griddle onto our dinner plates, my brother Nate folds them whole into his mouth.

> Celebration Cups (Prefilled Communion Glasses) 1) The Celebration Cup is perfectly designed to fit your communion trays. 2) Peel back the air-tight seal and eat the unleavened wafer. 3) Peel back the second seal and drink the juice. The plastic is left in the cup holders in the pew. $32.95 for a box of 100 (Religious-Supplies.com: 2008)

In a study of community in contemporary American culture, Robert N. Bellah likens discourses of individualism and consumerism to speakers' 'first language' and discourses of community and communalism to the 'second language' employed when the 'the language of the radically separate self does not seem adequate' (Bellah. et. al. 1985: 20, 154). Woman leaps in advertisement on side of passing bus: powerful thighs, Not Your Ordinary Church. Try it, you'll like it.

In my country this is how we say, 'Oh.'

In Rublev's fifteenth-century icon depicting the Holy Trinity, the father, son and spirit are three haloed, androgynous figures who share a table, wine and each other. Paradox of one and/in/as three gets resolved over supper. The divine beings live and dwell in each other, and there is a fourth place prepared for an invisible person, for you, the viewer, the communicant, to share: my dad tells me about this in the car driving to Madison at Christmas. Later he emails me a .jpeg

Performance Research 13(3), pp.184-187 © Taylor & Francis Ltd 2008
DOI: 10.1080/13528160902819547

of the icon. It's going to be projected on the back wall of the church. He's told me his sermon. It's good for him to have a picture of this second language he has been trying to imagine, like the relief you feel when an aspect of your own inner self has been articulated, summoned into existence by a novelist. The living, the dead, and the inchoate are invited to commune.

The word 'individual' is derived from 'indivisible.' 'One in substance or essence'

Each whole unbroken.

Each.

My mom had a hole in her side - it's funny, like Christ - for a whole year, on purpose, a fresh clean channel for her dammed liver.

Luckily, my dad is handy with gauze and iodine, spends a whole quarter of his job in hospitals, nursing homes, hospices and outpatient facilities, knitting socks suspended from four needles, waiting, comforting. Even old men, angry ones from the Council, want to hold his hand when they are scared and hurting. Take body and blood. Invite. The bed is the table or altar. Pass the bottle.

Invite.

The machines beep. Who has ever felt language was adequate.

A yellowjacket crawled under my shirt as I ate sliced rounds of banana and I ran crying to the women, quilting for Africa, holding rows of sharp pins in their mouths. Pins in aged lips so perilously pursed-I swallowed, the sting on my chest stung in my esophagus.

Names have been changed to preserve.
One in substance or essence.
All who have been baptized are invited to commune.

The first language has to do with surveys, slogans, mortgage rates, elected representatives. The church council will fix the roof by slashing missionary funding, hire stewardship consultants to increase giving, save on electricity by banning nightlights in the parsonage where I sleep and dream, drag a Ferris wheel into the parking lot at Easter and charge admission, institute a staff dress code, take out a colour ad in the yellow pages, institute a congregational dress code, resurface the parking lot, make do, cut salary if my dad can't fill an amphitheatre, can't get himself a thousand arms reaching out for more and more - winter nights and pancakes. My mom, my brother Nate and I around the table, framed in the window frame, our reflection in the lighted window. Pass the syrup.

Elect your Council representatives at the annual meeting: raise your hand. Count the arms in the air.

Afterwards, the annual feeding. Bring a dish to share.

Nothing's mine. We always invite someone we can't see. My brother ate my own baby teeth from out their velvet-lined box. Who am I without parts of me, who is he, devouring?

• *(left)* **'Crunch/Squirt' video still, from *Now memorable and convenient with prefilled single-use ready-blessed disposable communion cups*, 2008.**
• *(right)* **'Open your eyes you are my inspiration', video still, from *Vito Acconci is My Big Sister*, a Video By Nathan Anderson, 2007.**
(all images by Emily Anderson)

People who are mean to my dad hug me, put their arms around me. My body is not my own. I hug back. Carry sharp pins in your mouth if you want to join. We feed each other and through eating commune.

Like those recipes: 1/2 can love, 2 tsp. fairness, 'the magic formula, PLURALISM = MONISM' (Delueze and Guattari 1988: 23). We feed each other and through eating consume. All the figures' feet point to one another in Rublev's Trinity, make the shape of a mouth or a cup. Swallow. The viewer's invoked as ghost or honoured guest. Swallow. The folding table creaks mysteriously, the Jell-O trembles, shakes, it slithers from fork to plate. Is good with ham.

God could make me pregnant at any moment. Like Mary. Like right now.

• 'Halo', video still, from ***Body What Body: The traveling communion kit takes a bath*, 2008.**

'In Christology the term [*perichoresis*] is used to express the mutual interpenetration of different natures, divine and human, in the person of God-human-Christ' (Moltmann 2000: 113) A bee in my throat. A quilting bee. Knit our fingers together: here is the church, here is the steeple. Open all the doors, look at all the people. My body someone else's finger or open palm. My body more than a symbol less than a body.

Not your ordinary church. Powerful thighs. A leap of faith.

I don't *own* my body. I don't need to resurface it, bestow my proprietorial gaze onto it, be its shrewd councilman, put old breath in a young body, but I do. Ritual's a place at an uncanny table: when you sit down, like at Disney World, the animatronic figures come alive, unhinged, cough wingèd dust.

People who are mean to my dad and people who are nice to my dad bring vats of spaghetti alfredo and tuna noodles through the snow; because we are loved, I gain ten pounds when my mom has cancer. They bring brown bread and vegetable soup when my dad has a heart attack, tell him not to work so hard: he made the nurse wheel him down to the room of a dying parishioner to pray with her. Then the nurse wheeled him back up and she died. My body responds to compassion, you can see the church council in my hips and thighs. At church I stand tall, smile, blink at anyone I've forgotten who remembers me, hugs me.

You are surrounded.
You are invited.

After church we go to the mall. We have lunch with our family and friends in the food court. We are together, I get what I want. In the food court the individual may experience a feeling of 'illusory Sovereignty ... liberty may appear to be complete, for the aims and limits of the action and the means of execution seem to coincide, in perfect harmony' (Meszaros 1970: 258). We three kings. I have Happy Wok on a red tray. Dad has Arby's on a tan tray. Mom has Sbarro's on a green tray. Nate has Burger King on a blue tray.

Execute: family with table, chairs. Arms creaking with ghost. If this feels like a 'me-moir' it's because even though I've amicably separated from this and all churches and in fact live thousands of miles away I can't find a clear division between my past and the congregation's. I can think of a number of congregational controversies that have lived with my dad as long as I have. We sit under an umbrella and incandescents. We don't fight, we eat. We each get what we want. Enough fighting, enough food.

Please God don't make me pregnant. I'm only thirteen. Under my acolyte's robes, I'm invisible to myself, body somewhere else. I look in people's eyes when they slip their communion cups into the holes in my round silver tray. I know their secrets. I identify the alcoholics, taking the grape juice, pre-poured in tiny cups on crystal tray. I can tell whose knees hurt kneeling but who kneel and whose knees hurt kneeling but who stand. I smile, blink. The cups click into their circle-holes, I circle the altar, collecting cups, looking or not looking into eyes and mouths swallowing.

In the narrow sacristy we polish off the wine from the chalice while my mom washes the communion cups in soapy water. Dad and I pass the cup back and forth, I'm only thirteen, shouldn't be taking more than a sip, that's why it's fun, we're laughing: he doesn't have to shake any more hands now, he can take off his alb and stole and go to the mall where I'm not his arm or acolyte I'm materialistic, swinging an Esprit bag, demanding earrings at our umbrella table.

Red poinsettias on the altar at Christmas in honour of, mostly in memory of. Returning at Christmas I can only see the dead standing, sitting, standing like zombies, death decaying the teeth in their mouths. Sense voids orifices. Everyone old. Say OH! louder this time. Hold longer. Table's a narrow bed narrowing and to sit down together we have to love each other and get close or fall inside.

God can see me in the bathtub or locker room or everywhere or now. Like when mom and dad took me to see the dead woman in casket in front of the altar. Death, teachable death. We stood together. 'Don't be afraid. You can touch her.' I reached up into the casket, put my hand inside it, made death tangible. Her cheek was cold and pink, the colour of her glasses frames and her scalloped polyester dress. Someone must have changed her dress. Undressed her in some basement. Taken off her panties, put clean ones on for her burial. Moved her old pink legs and what was between them. Seen everything. Like God. I punched the red velvet altar rail and worlds of angels, sunlit motes surrounded me, whirling. I put my hand inside them.

Be my arm. Be like my arm, your lips where mine were on the cup, I'm going in circles, I'm eating mints and needles. A tray with an order of Happy Family. A tray with an order of pizza and Arby's Regular Roast Beef and a Burrito Supreme.

Make your mouth like mine, blink at me, invite me, come and sit between me.

REFERENCES

Bellah, Robert N., Madsen, Richard, Sullivan, William M., Swidler, Ann and Tipton, Steven (1985) *Habits of the Heart*, Berkeley: University of California Press.

Delueze, Gilles and Félix Guattari (1988) *A Thousand Plateaus*, trans. Brian Massumi, Minneapolis: University of Minnesota.

Meszaros, Istvan (1970) *Marx's Theory of Alienation*, London: Merlin Press.

Moltmann, Jürgen (2000) 'Perichoresis: An Old Magic Word for a New Trinitarian Theology', in M. D. Meeks (ed.) *Trinity, Community and Power*, Nashville: Kingswood Books pp. 111-73.

Religious-Supplies.Com (2008) 'Celebration Cups', advertisement online, <http://www.religious-supplies.com/allinonecelebrationcups-boxof100.aspx>, (accessed 26 December 2008).

[2] H P Lovecraft, *Through the Gates of the Silver Key in The Dream Quest of the Unknown Kadath* (Ballantine, 1970), pp.217-218

'... every figure of space is but the result of the intersection by a plane of some corresponding figure of one more dimension... and so on up to the dizzy and reachless heights of archetypal infinity. The world of men and of the gods of men is merely an infinitesimal phase of an infinitesimal thing.'[2]

Performance Research: Congregation

Notes on Contributors

THE EDITORS

Ric Allsopp is a founding editor of *Performance Research*. He is currently Senior Research Fellow in the Department of Contemporary Arts at Manchester Metropolitan University, UK, and Visiting Professor at the University of the Arts, Berlin.

Richard Gough is general editor of *Performance Research*, a Professor at Aberystwyth University, Wales, and Artistic Director for the Centre for Performance Research (CPR). He has curated and organized numerous conference and workshop events over the last 30 years as well as directing and lecturing internationally.

ISSUE EDITOR

Claire MacDonald is a founding editor of *Performance Research*. She is Director of the International Centre for Fine Art Research at the University of the Arts London, and a Contributing Editor to *Performing Arts Journal*, New York. Her research has two aspects: an interdisciplinary critical practice focusing on art, language and performance, and a related creative practice that addresses the questions of what language can do within the conditions and parameters of literary and performance form. Between 1998 and 2005 she lived in Washington DC, and now divides her time between the UK and Greece, where she writes about contemporary Greek art and performance. Her most recent play, *Correspondence*, was directed for Menagerie Theatre by Patrick Morris. She holds a PhD in Critical and Creative Writing from the University of East Anglia and recently completed her first novel. She is currently writing on performance collaboration and co-curating, with Claire Hind, *York - New York*, a series of exchange commissions for writing artists.

CONTRIBUTORS

Emily Anderson's writing has appeared in numerous journals, including *McSweeney's Quarterly Concern* and the *Denver Quarterly*. While completing her MFA at the School of the Art Institute of Chicago, she performed or exhibited work at various Chicago venues, including the Museum of Contemporary Art, the Gene Siskel Film Center and XTV. She lives in Madrid.

Sharon Aronson-Lehavi is Assistant Professor of theatre studies at the Department of Comparative Literature, Bar Ilan University. She writes about medieval performance, contemporary biblical theatre and Israeli theatre and performance.

Rina Arya is a Senior Lecturer at the University of Chester and lectures in art history and theory. She is interested in the relationship between art and spirituality and the possibility of thinking about the sacred in secular culture.

Carol Becker is Professor of the Arts and Dean of the School of the Arts at Columbia University. She is the author of several books and numerous articles. Her books include: *The Invisible Drama: Women and the Anxiety of Change* (published in seven languages); *Zones of Contention: Essays on Art, Institutions and Anxiety*; *Surpassing the Spectacle: Global Transformations and the Changing Politics of Art*. She is also the editor of *The Subversive Imagination: Essays on Art, Artists and Social Responsibility*. Her newest book *Thinking in Place: Art, Action and Cultural Production* is published in autumn 2008.

Dominika Bennacer is a PhD candidate in the Department of Performance Studies at New York University. She has a background in eastern European experimental theatre and is the Associate Curator of the *Year of Grotowski / New York* festival. Her research areas include immigration, activism, quotidian performances of identity and embodied practices in Islamic orthopraxy.

Claire Maria Chambers Blackstock holds an MA in Transforming Spirituality from Seattle University, and is finishing her PhD in Performance Studies at the University of California, Davis. Her dissertation, 'Acting on Faith: Theology, Theatricality and the Performance of Belief', examines theology as a function for understanding performance.

Peter Civetta is a Postdoctoral Fellow with the Alice Kaplan Institute for the Humanities at Northwestern University, where he is undertaking a research project exploring the role and impact of religion in the 2008 presidential election cycle. He received his PhD in Theatre Studies from Cornell University in 2004.

Thomas Crombez studied philosophy at the University of Brussels and theatre studies at the University of Antwerp. He obtained his doctorate in 2006 with a dissertation on 'The Antitheatre of Antonin Artaud'. His current research project on Flemish mass theatre during the interbellum period is supported by the Research Foundation-Flanders (FWO). More information on his publications and research interests may be found at www.zombrec.be.

Ann David is a Senior Lecturer in Dance Studies in the School of Arts at Roehampton University where she teaches on both undergraduate and postgraduate

Performance Research 13(3), pp.189-190
DOI: 10.1080/13528160902860590

courses, focusing on Dance Anthropology and South Asian studies. She is a Research Fellow on an international research project entitled 'The Religious Lives of Ethnic and Immigrant Minorities: A Transnational Perspective', based in London, Kuala Lumpur and Johannesburg. Ann has spent several years researching cultural and religious practices in several of the Saivite temples in Greater London and has training in Sanskrit and the dance form of Bharatanatyam. She has presented at many international conferences and has publications in leading academic journals.

Malcolm Floyd trained as a musician at London, Exeter and Oxford universities, then taught and lectured in Kenya for seven years. He has been at the University of Winchester since 1993 and has run courses in World Musics, Music Education and the MA in Performing Arts. He is currently Principal Lecturer in Performing Arts, Chair of the Faculty of Arts Ethics Committee, Acting Head of Research and Knowledge Exchange, He has postgraduate qualifications in medieval music, Maasai music and education, and creative writing.

Nissim Gal is Assistant Professor of Modern and Contemporary Art at the Department of Art History, University of Haifa. He is the author of *'Ilana Salama Ortar - La plage tranquille'* (Montpellier) and of the forthcoming *Portrait of the Artist as Interior Design* (Tel-Aviv).

Romana Huk is an Associate Professor of English at the University of Notre Dame. Her books include *Assembling Alternatives: Reading Postmodern Poetries Transnationally* (Wesleyan University Press, 2003), *Stevie Smith: Between the Lines* (Palgrave, 2005) and a forthcoming study of avant-garde poetries, postwar theologies and postmodern philosophy / sociolinguistic theory.

Megan Macdonald is a temporary lecturer in Drama, Theatre and Performance Studies at Queen Mary, University of London. Her PhD project suggests that analyses of spiritual performance can account for the performative qualities of practices that both instantiate and produce belief instead of focusing on representative meaning.

Deborah Middleton is Head of Drama, Theatre and Performance at the University of Huddersfield. She edited the English translation of Nicolás Núñez's *Anthropocosmic Theatre* (Harwood, 1996) and is an authorized practitioner of Núñez's training form, *Citlalmina*. Her current research investigates concepts of mindfulness in creative states of being.

Lara D. Nielsen is an Assistant Professor in Theatre and Dance at Macalester College, Minnesota, where she teaches courses in Performance Theory and Theatre History. She completed her MA in Comparative Literature at the University of Minnesota and her PhD in Performance Studies at New York University.

Kirsten Norrie is a writer, musician and live artist. She has performed internationally in Greenland, Arizona, Spain, the Netherlands, Germany, Vietnam and the U.K. After completing a masters degree in performance and writing at The Ruskin School of Drawing and Fine Art, she was given a Sciart award from the Wellcome Trust to make work based on the specific mass of her internal organs in relation to cultural metaphor. She has written for *Art Monthly* and is currently releasing her first album.

Norman Shaw grew up in the Highlands of Scotland. He is an artist who is currently a lecturer in Fine Art at the University of Dundee, and he is a member of the 'Window to the West' research team, who are exploring aspects of the visual in the Gaelic culture of Scotland. His current research project, entitled Nemeton, is a collaborative intermedial exploration of the contemporary Highland landscape. His work involves image, text and sound.

Kim Skjoldager-Nielsen is currently a part-time lecturer in the Department of Theatre and Performance Studies, University of Copenhagen, and the Department of Performance Design, Roskilde University Centre, teaching theory of rituals and performance theory. Topics of publication and research include dramaturgical processes of performance theatre, interactive theatre, and religion, rituals, theatre.

Elanor Stannage works as a theatre practitioner in various communities throughout Yorkshire. Her research into the healing performance informs the collaboration between herself and Nathan Walker.

Susan Tenneriello is Assistant Professor of theatre in the Fine and Performing Arts Department at Baruch College, CUNY. She writes extensively on theatre, dance and visual art, pursuing interdisciplinary work in cultural aesthetics. Her current book project examines popular narratives and the idea of nationhood in nineteenth-century American spectacle entertainments.

Nathan Walker is a homotextual artist working with photographic objects, actions and found items, including words. His research into the notion of care in contemporary art is explored with Elanor Stannage in their performance *Strong Hours Placed.* www.stronghoursplaced.co.uk

Samuel Wells is Dean of Duke University Chapel and Research Professor of Christian Ethics at Duke Divinity School, Durham, North Carolina. He has written numerous books and articles on Christian social ethics. His most recent book, co-edited with Sarah Coakley, is *Praying for England: The Heart of the Church* (Continuum, 2008).

For Product Safety Concerns and Information please contact our EU representative GPSR@taylorandfrancis.com Taylor & Francis Verlag GmbH, Kaufingerstraße 24, 80331 München, Germany

Batch number: 10731488

Printed by Printforce, the Netherlands